Traditional Math

An Effective Strategy that Teachers Feel Guilty Using

Barry Garelick & J. R. Wilson

John Catt

First published 2022

by John Catt Educational Ltd,
15 Riduna Park, Station Road,
Melton, Woodbridge IP12 1QT

Tel: +44 (0) 1394 389850
Email: enquiries@johncatt.com
Website: www.johncatt.com

ISBN: 978 1 915261 54 0

Set and designed by John Catt Educational Limited

To teachers, parents, and last but not least our students, who have served as the final arbiters on how best to teach math.

AUTHORS

J.R. Wilson has thirty plus years' experience working in public education as an elementary classroom teacher, middle and high school math teacher, state department of education curriculum consultant, regional educational service agency staff development coordinator, and elementary principal. As a team member he has been involved in writing science and math standards. He has served on the Executive Committee of a math advocacy group in Washington State called Where's the Math? He has participated in and coordinated reviews of the draft Common Core Math Standards, and state standards. Even though retired now, J.R. continues to work on education issues along with other volunteer work. He and his wife are residents of the Pacific Northwest.

Barry Garelick taught 7th and 8th grade math for ten years as a second career after retiring from the federal government where he worked in environmental protection. He majored in math at University of Michigan. He is the author of several books on math education, including his most recent, *Out on Good Behavior: Teaching math while looking over your shoulder*, published by John Catt. He and his wife reside in the central coast area of California.

CONTENTS

FOREWORD: GOOD EVIDENCE-INFORMED EDUCATION IS THE NEW PROGRESSIVE!

Houston, we have a problem.

Math is essential for everyone. Without being able to "do the math," you will not be able to understand baseball scores and its statistics (games behind, earned run average, batting average), what the dosage of medicine to give for yourself or others (this many grams per kilogram weight), whether a bargain is really a bargain or if you're being fooled by either advertised 'twofer' prices (two small bottles cost more but have less contents than one large bottle) or shrinkflation (instead of raising prices, you shrink the size of the product), and so forth.

This means that math has to be taught in a way that ensures that students learn math as well as possible. In the past two or three decades, we've seen a change. Until 1990, U.S. students were taught to solve math problems by applying memorized facts and procedures—which are significant artifacts of what we'll call traditional math education. Since 1990, both the United States and many other countries around the world have de-emphasized this traditional approach and stressed applying reasoning strategies over, and often instead of, acquiring knowledge and skills; let's call this the progressive math education.

This progressive approach was first developed by the Freudenthal Institute in the Netherlands in 1971 and was given the name Realistic Mathematics Education (RME), a name still in use today. According to

Yuanita, Zulnaidi, and Zakaria (2018), RME begins with the students' daily life-experiences. These experiences are then transformed into a model of "reality" that are referred to as a "context". The goal is to develop long-term mathematical understanding that students can make sense of. RME does this through "a mathematical vertical process before turning it into a formal system" (p. 1). Put into plain English, the RME method teaches math concepts before teaching the fundamental procedures students need to achieve the goal of understanding. In RME, the teacher is a facilitator (also known as "guide on the side")—who helps students solve problems in the context of mathematical concepts, which its advocates assume will develop students' mathematical representation and understanding.

This change in approach to teaching mathematics was based on the untested assumption that our brains could reason in math via concepts as effectively as they do in applying facts and algorithms that are well-organized in memory. It was therefore assumed that by making math realistic, motivational, and fun, kids would and could learn better. Unfortunately, this assumption is mistaken. Research in cognitive science has shown that "working memory"—where information processing and problem solving occurs—is extremely limited. Because of this, reasoning that does not rely on memorized procedures and facts generally fails when students are solving complex problems.

The sowing of this progressive RME approach to math has reaped some shocking results. One example of this is the International Test of Numeracy Skills by the Organisation for Economic Co-operation and Development (OECD), for citizens aged 16 to 34. In 2012, among 22 developed nations, the United States ranked twenty-second and last. (Goodman et al., 2015). Why?

Though the evidence is correlational, the K–12 state math standards adopted since 1990 in the United States have changed how math is taught. With the rise in availability and use of calculators, understanding and mathematical reasoning became emphasized, and memorization of facts and procedures were de-emphasized. The un-noticed, unstated, untested, and optimistic assumption was made that we can reason about things without domain-specific knowledge and skills and that our brains, when solving problems, could apply new, looked-up, mentally calculated at that moment, and "calculator calculated" facts on a "just-in-time" basis as easily as if applying previously learned and mastered material.

Scientists who study how the brain functions when we solve problems have shown that the opposite is true, however. Working memory—where information is processed and problems are solved—has an exceptional ability to apply well-practiced math procedures (i.e., algorithms) that in turn rely on well-memorized facts (Hartman et al., in preparation). Research in cognitive science has identified especially efficient and effective strategies (e.g., spaced practice, interleaving, retrieval practice, overlearning) to move math facts and procedures into long-term memory so that they can be easily and accurately retrieved when necessary. In states that have aligned math standards with recommendations of cognitive research, scores on international tests have equaled and even exceeded high-ranking Japan (Hartman & Nelson, 2016; National Center for Education Statistics, 2015).

This book will, hopefully, provide an arsenal of tools and techniques to break through this downward spiral in teaching and learning math. First off, it breaks through the myth and straw man that explicit instruction (Rosenshine, 2010, 2012) is just boring chalk-and-talk rote learning of facts that cannot be applied when needed. It also breaks with the misrepresentation that traditional math teaches kids to work as automatons without understanding what they do.

Traditional Is the New Progressive

This so-called progressive math education has been anchored in education policy, schools, teachers, teacher training colleges and courses, education trade unions, politics (and legislation) at all levels, and university educational science faculties for more than 65 years (Kirschner, 2009). It has even become a part of the education press (e.g., TES: Times Educational Supplement) in the United Kingdom.

In other words, progressive education has, slowly but surely, become the norm with "new" progressive education innovations stacked on top of older innovations. And what has been the result? A continuous deterioration in pupil performance (i.e., learning), accompanied by an increase in behavior problems in the classroom (with excesses, e.g., verbal and physical violence between pupils and against teachers).

Interestingly, teachers were blamed for these failures. Teachers did not properly implement progressive education. Teachers didn't motivate the children properly. Teachers couldn't control the class. Another often-used scapegoat was the environment because "what could you expect from children of low socioeconomic classes?" This last aspect is what Robert Peal (2014) calls "the soft bigotry of low expectations" in his book *Progressively Worse: The Burden of Bad Ideas in British Schools.*

But something that has been in place, used, and *failed* for more than 50 years has no right to call itself progressive. Richard Mayer (2004) spoke of the "three-strike rule" when it comes to innovations that continually rebrand themselves and then prove to be just as ineffective as their predecessor. In other words, this so-called progressive education is actually traditional now, and as a result, we can conclude that education

- that is first aimed at acquiring knowledge, after which that knowledge is used for the acquisition of skills and attitudes,
- where teaching is based on a scientific evidence-informed basis,
- where teacher training courses and teacher training colleges are given the time to, as teachers teach how children learn and which learning strategies can and should best be used for this,
- where the teacher is a teacher again and not a guide or facilitator,
- where children from the "lesser" backgrounds are challenged instead of tolerated, and so on…

… is actually new, very innovative, and progressive.

Thus, good evidence-informed math education, as the authors of this book present here, is the new progressive math education!

Paul Kirschner

Emeritus Professor of Educational Psychology—Open University of the Netherlands

Guest Professor—Thomas More University of Applied Sciences

References

Goodman, M., Sands, A., & Coley, R. (2015). *America's skills challenge: Millennials and the future*. Educational Testing Service. https://www.ets.org/s/research/30079/asc-millennials-and-the-future.pdf

Hartman, J., & Nelson E. (2016). Automaticity in computation and student success in introductory physical science courses. http://arxiv.org/abs/1608.05006

Hartman, J., Nelson, E., & Kirschner (in preparation). Designing math standards in agreement with science.

Kirschner, P. A. (2009). Epistemology or pedagogy, that is the question. In S. Tobias & T. M. Duffy. *Constructivist instruction: Success or failure?* (pp. 144–157). Routledge. https://lexiconic.net/pedagogy/epist.pdf

Mayer, R. E. (2004). Should there be a three-strikes rule against pure discovery learning? The case for guided methods of instruction. *The American Psychologist*, 59(1), 14–19. https://doi.org/10.1037/0003-066X.59.1.14

National Center for Education Statistics. (2015). *Trends in International Mathematics and Sciences Study (TIMSS)*. U.S. Department of Education. http://nces.ed.gov/timss/benchmark.asp

Peal, R. (2014). *Progressively worse: The burden of bad ideas in British schools*. Civitas. https://civitas.org.uk/pdf/ProgressivelyWorsePeal.pdf

Rosenshine, B. (2010). *Principles of instruction*. International Academy of Education, UNESCO, International Bureau of Education. http://www.ibe.unesco.org/fileadmin/user_upload/Publications/Educational_Practices/EdPractices_21.pdf

Rosenshine, B. (2012) Principles of instruction: Research-based strategies that all teachers should know. *American Educator*. http://www.aft.org/pdfs/americaneducator/spring2012/Rosenshine.pdf

Yuanita, P., Zulnaidi, H., & Zakaria, E. (2018). The effectiveness of Realistic Mathematics Education approach: The role of mathematical representation as mediator between mathematical belief and problem solving. *PloS One, 13*(9), e0204847. https://doi.org/10.1371/journal.pone.0204847

INTRODUCTION BY BARRY GARELICK

I believe strongly in how math should be taught, which happens to be what has come to be known as the traditional method. Nevertheless, when I teach math in this mode, I feel vaguely guilty, as if I am doing something against the rules and perhaps even wrong.

I have heard from other teachers who identify and empathize with this. (For the record, J. R. Wilson, my coauthor, is not one of them, and I'm sure there are others.) The term "traditional math" is fraught with images and mischaracterizations:

- The teacher stands at the front of the room and lectures nonstop for the duration of the class, students learn all procedures and problem-solving methods by rote, and no background on the conceptual underpinnings of the same are presented.
- Students are viewed as "doing math" but not "knowing math" and have no conceptual understanding of what they are doing.
- Topics are presented in isolated fashion with no connections to other topics. Word problems are dull and uninteresting, and students do not feel any desire to try to solve them.

Despite the pervasive criticism of traditionally taught math, there are teachers who continue to teach math in this manner. They do so because they believe it incorporates pedagogical methods that have been proven to be effective, such as explicit instruction, worked examples, and scaffolded problems.

I taught seventh- and eighth-grade math as a second career, spanning about 11 years after retiring from a career in environmental protection. I majored in math in college, used it in my first career, and realized I wanted to teach in my retirement. I have written articles about math education over the years, and chose to teach it the way I was taught: that is, via explicit instruction with worked and scaffolded examples.

This book provides a glimpse of what explicit instruction looks like in the classroom for K–8. The book seeks to inform readers of our approaches to traditionally taught math. This has two purposes: 1) For approaches that are similar to what you may already practice, it provides some assurance that you are not alone, and 2) it may give you some new ideas.

Part I is written by J.R. Wilson and provides information about many important key topics in math for K–6. Over the years since entering the teaching profession in the 1970s, J.R. found strategies and approaches to teaching math that worked as he helped students understand and be successful with math. He had no intention of writing a math textbook; rather, he shares his approach and strategies as he addresses important topics and skills that provide students with a foundation for success in math as they advance through the grades and life. His talk-through of rounding numbers and long division provides a glimpse of how he talks through problems during scratch-paper time.

Part II covers regular seventh-grade math (Math 7) and Accelerated Math 7. Accelerated Math 7 covers all the topics in Math 7 as well as those in eighth-grade math (Math 8). A sample of topics and lessons from both courses are in Part II. Part III includes a sample of topics and lessons for eighth-grade algebra.

The example lessons in Parts II and III illustrate what explicit instruction looks like in the classroom. They show how particular topics and procedures are explained, the types of worked examples used, and how previously covered topics are kept fresh so that students remember them when they come up again. Most importantly, the lessons show how traditional math teaching addresses not only the procedures, but also the concepts behind them.

I recently retired from my teaching career, so my descriptions of lessons and explicit teaching are taken from very recent memories of my classroom experience. J.R.'s descriptions go back further in time and embody the perspective of his teaching experience from a time that embraced those methods. Despite experiencing our teaching in different times, we are both oriented to traditional math teaching. It wasn't because we were both taught that way, as some may believe, but because that method worked for us, and we have seen it work for our students. It is efficient, effective, straightforward, and it helped our students develop mathematical reasoning, understanding, and confidence. Most importantly, it helped them to be successful.

Textbooks

Textbooks are necessary and useful because they contain a logical sequence of topics and a breakdown of what gets taught within each one. Unfortunately, many of today's textbooks are poorly written with very little explanation of a procedure, and frequently two or three subtopics are embedded in a single lesson. Our approach in this book is to provide workarounds to these shortcomings. This frequently involves taking more than one day to teach a book's lesson. It also involves supplementing material that is in the textbook with material in other books, which are referred to throughout.

For seventh-grade textbooks, I usually had to reorder the sequence of topics for a more logical flow, add material within each topic area, and provide problems taken from other books. It does not involve designing a new curriculum from scratch. For both seventh- and eighth-grades, I had sometimes crafted homework worksheets using problems from a variety of sources, and I ordered the problems so that they started with easy problems and progressed in difficulty. Present-day textbooks I have worked with often start with more difficult problems, as well as problems for which no examples or discussion have been provided.

For eighth-grade algebra, I used a 1962 algebra book by Dolciani et al. Although this book is very good, its availability on the Internet has decreased to the point that it is now very expensive. Also, some of the topics are presented very formally—indicative of the set theoretical approach that was in vogue in the 1960s New Math era when the book was written.

The following is a list of the books I draw upon, which are also mentioned throughout. (Publication information for these books is in the References at the end of Part III.)

- Brown, Smith, and Dolciani (1990) *Basic Algebra*
- Mary Dolciani et al. (1962) *Modern Algebra: Structure and Method*
- JUMP Math (2015) *Assessment and Practice, 7.1 and 7.2*
- Paul Foerster (1990) *Algebra 1: Expressions, Equations and Applications*
- Singapore Math (2003) *Primary Math series*

Many of the problems included in the example Warm-Ups come from Singapore's *Primary Math* series and Dolciani's *Modern Algebra* textbook.

CLASSROOM ROUTINES

Warm-Ups. These are four or five problems that students work on in the first 5–10 minutes of class. Some of the problems are from previous lessons to keep old material fresh. Others are from material that has just been covered. Still others may be problems that lead into the day's lesson. The latter types of problems help set the stage for the day's lesson. I provide hints and guidance while students are working on the problems, and then I go through the answers after time is up.

Going Over Homework Problems. I provide the students with answers to the homework problems so the students have already checked their work. Therefore, I spend this time going over problems that they find difficult—usually three or four problems. I may have a student who has done a problem correctly explain it; otherwise, I explain.

The Lesson and Starting on Homework. The lesson examples in Parts II and III of this book provide information on what is talked about and how it is delivered explicitly. The pattern followed is the "I do, We do, You do" technique (which J.R. describes in Part I, Chapter 1). This technique employs worked examples that are part and parcel to the instruction, followed by examples that students must do independently, with guidance from the teacher as needed.

I leave enough time (approximately 15 minutes) at the end of the lesson for students to start working on their assigned homework problems. This allows me to answer questions and provide help and guidance. It also works to prevent the situation of students not knowing how to do the problems when they get home.

Some students will work with others on the problems, but I limit this to no more than two people. I monitor those working with a partner to make sure they are on track. Starting the homework problems in class allows for practice and the learning that comes from it. Most importantly, I try to stem struggling with a problem and aim to have students be successful. My guiding principle is that struggling to learn the breaststroke is not the same as struggling to keep from drowning. The latter doesn't teach you how to swim.

When Students Make Mistakes. I frequently hear that it is not a good idea to tell a student that they are wrong, even though some advocates say that making mistakes should be a goal of teaching math. It is not a goal for me, but mistakes will happen. I don't shy away from telling a student that they are incorrect. Some people use mini-whiteboards that students write their answers on and then hold up.

I have several methods for dealing with mistakes. First and perhaps most important is what happens when I make a mistake—which I do. I tell the class the first day that if a student catches a mistake that I've made, they will receive a package of Goldfish crackers. Although showing students that even teachers make mistakes helps them see that it is part of doing and learning math, this doesn't stop the shyness and fear of being wrong in front of one's peers.

If a student responds to a question I ask with the wrong answer, I might say, "Not what I got." If I can see what the mistake was, I will point it out. For example, I might say, "Oh, it looks like you multiplied instead of divided," or whatever the mistake happens to be. I go around the room and call on others. If there are many mistakes, I make a game out of it, writing down the answers on the board until the correct answer comes up. I make a point of coming back to the students who made a mistake so they can try again.

In general, students in middle school can be quite guarded about mistakes. One technique that has worked well is to ask students to work on a problem in their notebooks. I walk around and will offer help or guidance for the weaker students, and once I see they have solved the problem correctly, I have them present the answer and how they worked on it to the class. Knowing they have it right fills them with the needed confidence to do this. (Some teachers rely exclusively on mini-whiteboards. I prefer notebooks for the reasons stated above, but also because my eyesight is not all that great, so mini-whiteboards can be hard for me to read.)

I also encourage students to try to solve challenging problems by offering a Kit Kat bar for the first student to come up with the correct answer—and show the class how they did it. This has proven to be unusually effective.

Checking for Understanding

Much is written about "checking for understanding" and "formative assessment." Without getting too far into what constitutes understanding, in this context it means absorbing what has just been taught and applying the basic procedure or concept to answer the teacher's questions. Teaching is a combination of providing instructions and asking questions. Questions may address material just stated, and other times they require students to make a conjecture. This last point is done to provide a segue to the next part of the explanation as well as to check on whether and to what extent students are following instruction. Students are kept on their toes knowing they could be called on. It also helps teachers know if their explanation is working as is or needs to be altered.

Instructions and questions take many forms. The Warm-Ups provide a check on material presented in past lessons, as well as front-loading material covered in the upcoming lesson. Worked examples are a source of checking and reinforcing instruction. Extra credit and challenge problems provide a way to evaluate ability levels and identify students who may need more of a challenge. Tests and quizzes also provide information on students' understanding, and to that end can serve as a formative assessment even if a test is meant to be summative. They provide a window into what areas students need more practice and instruction. In short, there is no one way to check for understanding; methods will differ for every teacher.

ONE LAST WORD

The book is designed so that you can pick and choose which topics or lessons you want to read about. It is a reference to what traditional teaching is—and isn't. In Parts II and III of the book, in which examples of lessons are provided, I have depicted some of the reactions and responses of students to the teacher's questions that I feel are realistic and only slightly exaggerated. It was not my intent to show that students do not try hard—quite the opposite. When encountering a topic for the first time, new information can be overwhelming, and students can experience various degrees of "brain freeze." To prevent this, I present new information incrementally, but even so, students are going through the balancing act of incorporating new information on top of what they have learned previously. I try to represent when it becomes too much, as well as show how to get beyond this inertia. We hope you enjoy this book, and we wish all teachers well in their teaching.

PART I
GRADES K–6

The topics in this part include the foundational areas of math that are built upon in algebra. Instruction and practice help students develop math skills. This development, or skill building, takes place gradually over time. Often, new skills build on previously learned skills. In Chapter 2 of this part, a Numbers and Operations Table is presented to serve as a guide for teaching skills and concepts at each grade. As in Parts II and III, Part I does not include geometry or statistics.

The topic-oriented chapters (Chapters 3–8) present the overall sequence of topics, though it isn't broken down specifically by grade. It is meant to give an overview of what aspects are covered and how they are presented. In keeping with this look and feel of traditional teaching, each chapter also includes—under the heading "Just for Fun"—math jokes I have told students. Such jokes have been met with both laughter and groans—mostly the latter. I hope you enjoy them—or try to at any rate.

CHAPTER 1. GETTING STARTED

Intro and an Oversize Smidgeon of My Background

I started out as a primary grade teacher and ended up teaching high school math as a long-term substitute. After teaching primary grades for a few years, I went to graduate school and earned an M.Ed. and Ed.S. in administration. I worked for a few years as a curriculum specialist at a state department of education and then as a staff development coordinator for a regional educational service agency serving 11 school districts. I had a goal of becoming an elementary principal. I achieved that goal for five years in two different positions—one as principal of two elementary schools concurrently and then as a teaching principal. I then gravitated back to the classroom, where my heart was. At some point, I taught all subjects at every grade in elementary school, kindergarten through sixth. My elementary teaching was capped off with eight years teaching sixth grade. While teaching sixth grade, I realized I enjoyed teaching math more than any other subject, so I stepped out of the elementary classroom. I obtained a middle school math endorsement and began substitute teaching in middle and high school math classes. Several times I was called on for long-term high school sub gigs teaching first- and second-year algebra, geometry, and remedial algebra. Some of those gigs lasted a semester or more.

Having taught every grade kindergarten through sixth, I have a fair sense of what should and can successfully be taught at each grade level. I draw heavily on my elementary teaching experience in what I share here. It is important to keep in mind that what I discuss is what I found worked for me and helped my students meet with success. My approach and strategies may work for some teachers, but not for everyone. Each teacher needs to find what works for them, to help their students be successful in math…and have fun.

My sixth-grade experience was in an elementary school where I was responsible for all subjects. This may have provided me with greater flexibility, at least timewise, than middle school teachers who may only teach sixth-grade math.

Success and Fun

Each year, in my sixth-grade classes about 25% of the entering students would be proficient with their multiplication facts. By early December, 75% would be proficient. In my observation, this roughly corresponded to student attitudes toward math. At the beginning of the year, most of my students hated math and would have been happy, cheering loudly, if math class was skipped for the day. As the school year progressed, a gradual shift took place to the point where most students cheered when I would announce it was time for math. What contributed to this shift? I doubt there is any one factor, but I think how math class was conducted had many contributing factors. As students began being more successful with math, it became more fun for them. Having fun and being successful are important elements. It also helped that students could tell I enjoyed math and was helping them become successful with it.

It's possible my attitude and enjoyment would rub off onto some of my students. I remember running into a former student on a visit I made to one of the junior highs my students would move on to. The former student excitedly told me that I made math come alive and be fun for her by having numbers talk to each other. A little numerical personification reached at least one student.

I feel I helped most of my sixth-grade students make a shift to like and enjoy math. Many would even express that it was their favorite subject. Not for lack of trying, there were always students I could never reach. Although there were any number of reasons these students weren't successful, they were students who put forth little to no effort to pay attention or perform during math class or with assigned work. For many students, it is difficult to be successful without effort. For most students, things are more fun when they are successful.

Finding My Way by Using I-We-You

Each teacher needs to find what works for them. Early in my career, I found something that worked well for me. It ended up not just being something in my so-called bag of tricks, but it was something I used almost every day in class for the duration of my career. Over the years, I would tweak it here and there to help more students be successful.

What is this something, and how did I find it? I, along with many others, call it "I do, We do, You do," or I-We-You for short. In more recent years, I have seen offerings for staff development called Gradual Release of Responsibility that incorporate or are built upon this method. I really can't say I found, discovered, or developed this method. It was a gift from one of the best principals I had the pleasure and opportunity to work with. At the end of a post-observation conference, he briefly explained this method to me and suggested I give it a try. It made sense to me, and once I tried it, I continued to use this method. In what follows, I explain how I use this method. If you are a teacher and this is new to you, I suggest you search for more information online and give it a try, tweaking the method to suit yourself. Try different things to find what works for you.

Phase 1: I Do. This phase introduces what we will be working on. I explicitly work on examples as I explain each step of the problem. I cycle through this step working on as many examples as necessary until it seems students are ready to try one on their own. After a few examples, asking questions such as, "What should I do next?" can serve as a quick check on student understanding. During this time, students shouldn't have a pencil in their hand and should be paying attention by listening and watching.

Phase 2: We Do. This phase is guided practice. In this phase, I put up a problem just like what I worked on in Phase 1 and ask the students to try working on the problem on their own. While they are working, I walk around the room, seeing how students are doing. I might comment or give pointers as I view student work. This is also an opportunity to provide some individual help as needed. I also make a mental note of types of errors I observe students making. When it appears most students have done what they are able, I ask them to put their pencils down and pay attention. I then work through the problem for them and explain it step by step. This gives students a chance to self-assess by checking their own work. I may also point out common errors people make and suggest ways to avoid them. When I work and explain the problem, it is another opportunity to pose questions to the class such as, "What do I do next?" and "Why would I do that?" I also ask if anyone has any questions and address these questions. We cycle through this process until it appears everyone is getting it. Then, we move on to Phase 3. In Phase 2, some teachers may ask students to work as partners. Even if students work with a partner, the teacher should make sure each student is able to be successful and that one student isn't carrying their partner. Some teachers may want to ask students to use individual slate boards or dry erase lapboards and raise them to show when they complete the problem. Those items were usually never available for me, so I grew accustomed to having students use scratch paper and would ask students to raise their hand as a signal to me that they completed a problem.

Phase 3: You Do. This phase is independent practice. At the end of Phase 2, I assign a set of practice problems for the students to work independently and provide in-class work time sufficient for students to complete the set. We then check the assignment at the beginning of math the following day. Most students would realize that if they used their time well, they would finish their work in class. Some wouldn't use their work time well and would then have the remainder of their assignment as homework. Those students who didn't use their class work time well had a tendency to struggle since it wasn't fresh in their mind when they got home and they might not have anyone available to ask for help. Student use, or lack of, class work time, led to two types of comments from parents. Parents of students who used their time well and finished the assignment in class would comment that their students never had any homework. I would explain to them that it was a good thing since their student used their class work time well. I would let parents know, they can always ask and expect their student to bring their completed assignment home to show them. Some parents of students who didn't use their class work time well, might comment or complain their student either had too much homework or they didn't understand the work. I would let parents know adequate work time was being provided in class, with help available, and ask them to encourage their students to make better use of the work time.

When I started using this method, I used an overhead projector. Later, I used a document camera and a projector. Although these are nice to use, if they aren't available, a chalkboard or whiteboard will also work.

Scratch-Paper Time. I always had a bin of scratch paper in the classroom. Students knew it was for them to use when they needed it, and they could get some at any time. I would explain the "I do, We do, You do" phases to my classes early in the year. After that, I would refer to it as "scratch-paper time." When I would announce, "It's scratch-paper time," students knew to clear their desks of everything except a pencil and some scratch paper and be ready to pay attention. It became a routine with clear expectations and an opportunity for students to comfortably strive for success.

Three things I would use the I-We-You strategy for were pre-teaching, prepping for current work, and backfilling deficiencies. The first was to introduce something new and pre-teach. If possible, I would try to cycle through Phases 1 and 2 with examples from one or two lessons ahead of our current work. That way, when we got to those problems in a few days, students were already familiar with them.

The second way was to prep students for the assignment they were going to get that day. Prior to this prep, we might have a quick review of the previous day's work. This review often would have already taken place, as we would check our work from the previous day's assignment. Since the current work would often build upon the previous day's work, a quick review and making the connection to how it relates is helpful.

The third way is to backfill deficiencies. There were always skills students should have coming into sixth grade that they didn't all have. I would use scratch-paper time to help bring everyone up to speed on those skills. These may have been skills that hadn't been taught, that were taught differently by different teachers, or that students just didn't understand for whatever reason.

Performance Inspection

In this section, I am referring to tests used in class to test what has been taught, rather than state-mandated assessments. These would either be tests I would make, ones accompanying the textbook being used, or district-made tests for units in our math program. Without dwelling on it and never spending an inordinate amount of time, I would orient my classes on test-taking skills.

I went over, explained, and practiced examples of every kind of problem that would be on one of our chapter tests during scratch-paper time. After one test, I remember hearing one of my students say to another that I had showed them how to do every problem on the test. Test what you teach!

The results of a test, for me, were as much an indicator of how well I had taught the material as they were an indicator of how well the students were doing. Results would let me know if we needed to spend more time on any specific skill as well as which students might need some individual help. Pop quizzes with three to five problems were also a good way to get a read on how students were doing.

Early each year, I would familiarize myself with the report card so I would know what kinds of things I needed to record grades for. When I taught sixth grade, there was always one item that troubled me. I had to provide a grade for each student on math reasoning. Every year, I would search to see if there was a clear definition for what that meant as well as a way to assess it. I never really found anything helpful, so I would identify some problems that I would use as indicators. I was always concerned and wanted to be prepared with something to show a parent to justify a reasoning grade if any parent asked what it was based on. Fortunately, no parent ever asked about this.

Math Splaining: Showing the Work

Do I expect students to explain how they get their answers on math problems? Absolutely, with almost each and every problem. How do I expect students to explain? An essay? Writing out strategies? Using sentence stems such as "The first step I took to get my answer was…" to get them started? Asking guiding questions?

Nope.

I ask students to explain how they get their answers by showing their work step by step using the language of math. Math talk and modeling takes place during scratch-paper time. I show and model for students how I expect them to show their work. As work progressed with a particular type of problem, I might begin to show students some shortcuts to take when showing their work. I would let them know that anyone should be able to look at their work and follow it from beginning to end and understand what they did to arrive at their answer. Seeing student work is helpful in at least two ways: 1) It can be an indicator of their understanding, and 2) if they arrive at an incorrect answer, it should be easy to go through the steps of how they worked the problem and find where an error was made.

Checking Work: Self-Assessment

At the beginning of math class, I would ask students to get out their assignment from the previous day along with a checking utensil. I would then provide the answers to the problems, usually by saying the answers out loud as I walked around the room, making note of who completed the assignment and who didn't. This required students to really pay attention and follow along as they checked their own work. They were to mark any problem they didn't get correct. When I finished giving out the answers, I would ask if anyone wanted me to repeat any answers or if there were any problems anyone wanted me to work through for them.

For most of my sixth-graders, checking their own paper was a shift from pleasing the teacher to assessing themselves. At first, many students would hurriedly try to erase a wrong answer and write the correct one on their paper without regard for why they had missed it. Eventually, students began trying to figure out why they missed a problem. This gave students immediate feedback on their work and the opportunity to get further explanation on problems they may miss. I didn't collect assignment papers on a regular basis. I might collect them on random days so students would know they needed to be prepared.

Once I started having students check their own work, I began to see more of them take responsibility for their learning. Rather than trying to please me as the teacher, more of them were striving to be successful with their math work. If they couldn't figure out why they missed a problem, they would ask me to work the problem for them so they could figure out where they made their error.

Separating the Wheat from the Chaff

It's important to know what's important for students to learn. Standards can be helpful as one guide for a teacher to use and may be even more helpful if the teacher is familiar with the standards for two, three, or more grades beyond the one they teach. Knowing the math and standards beyond the grade one teaches shows where things are going and can help one know what is important. Even better for an elementary teacher is being familiar and comfortable with algebra. The instructional focus should be on what is important for a student to be successful as their math instruction advances through the years.

I have observed an elementary teacher spend several weeks teaching about and having students work with Fibonacci numbers. Fibonacci numbers are interesting and can be fun for students to work with when exploring patterns. Is it important to spend a lot of instructional time working with this pattern? This pattern only gets brief mention in some calculus texts and is seldom mentioned in math programs through high school. Time would be better spent at ensuring students are fluent with multiplication facts and are able to successfully complete division problems. Fibonacci numbers: wheat or chaff?

Early in my career while I was teaching second grade, I came across a new math text that spent a lot of time on base eight. The idea behind this was to help students understand place value. Since it was in the book I had to use, I did address this but didn't spend too much time on it. I didn't see this as important. Since then, without teaching base eight, I didn't see students having difficulty with place value. Base eight did confuse students (and teachers and parents). In my life and career, I have never needed to use base eight. Although it might be used in computer science, I have not come across it being used or needed in algebra, geometry, or pre-calculus. Base eight: wheat or chaff?

A good math book will be filled with wheat and have little, if any, chaff. Unfortunately, publishers seem to have a greater interest in selling their books and programs than providing the best material for a solid math foundation for students. As a result, it is important for elementary teachers to know what's important, leave out or not spend much time on the unimportant (e.g., base eight), and add in important things a text or program leaves out (e.g., telling time).

Sophisticated Level of Understanding

I have heard others say a teacher needs to know and be familiar with the math two years beyond what they teach. For elementary teachers, I think it is beneficial to know math through at least first-year algebra. When teachers know what math students will experience in the coming years, they are not only in a better position to prepare their students but can separate the wheat from chaff.

It is important that teachers not only know the mathematics they plan to teach but also be able to explain it. That is the emphasis of the following quote from James R. Milgram (2005).

> *Second, we must ask whether or not basic subject knowledge of mathematics is sufficient to teach mathematics well. Increasingly, authorities such as Liping Ma have argued that to do an excellent job in the classroom, teachers must know elementary mathematics at a much more sophisticated level. For example, a teacher should know not simply that one can multiply by ten by "adding a zero" but be able to explain why this is true. A teacher should not simply know how to divide fractions, but why the rule for this operation is true. More generally, a well-qualified teacher should understand that mathematics is not a system of rules but a system of thought, that the rules make sense and can be explained. If we are to raise a generation of children who are prepared for an increasingly mathematically sophisticated world, we must explain the meaning of mathematics from the first instant.*

This makes a strong statement about teachers needing to be able to explain. This should not be mistaken as saying students should be able to explain. Students can be successful with math without being able to explain why or how what they are doing works. It is nice if they can explain. Often, when introducing and explaining something new, I will ask students to explain steps I do when working examples. I don't dwell on requiring students to explain everything.

Elementary teachers can improve their math knowledge and ability to explain in a number of ways. One way is to take an appropriate-level math class at a community college. By appropriate, I mean one that will be a refresher for the teacher and is likely to relate to the math the teacher is responsible for teaching. Calculus would not be such a course, but a basic mathematics or beginning algebra course might be ideal. A second way is to find some good solid older math textbooks, such as Saxon or Singapore, to use as reference in preparing math lessons. The idea is not necessarily to learn new material, but to learn the math to be taught at such a "sophisticated level" that the teacher can provide understandable explanations with confidence.

Importance of Understanding

This is good place to tell a story. It has implications for math instruction even though it is not a math story. While I was teaching fifth-grade social studies, one student provided me with an enlightening experience that to this day stands out in my mind. This student (I'll refer to him as Joey) was an eager and capable student. I noticed Joey struggle with social studies, and it puzzled me. He was a capable reader and did the reading in the book. He frequently raised his hand to answer questions that were asked of the class. However, when called on, he provided responses that didn't answer the question asked. I noticed the same thing with his written assignments. I made it a point to talk with Joey. Joey didn't realize the answers to the questions could be found in the book. He had never made that connection. He shared that he thought I accepted answers from students I liked and that I didn't accept answers from students I didn't like. He was certain I didn't like him. I showed him how to find answers in the text to the questions at the end of the chapter. The light bulb went on when he made the connection and understood acceptable answers weren't random and that he could find and provide a correct answer and show where it came from. Joey went on to be one of the top social studies students in the class, and his confidence had a great boost.

In some ways, this relates to math instruction. If a student makes consistent mistakes with the same kind of math problem, the teacher should be able to diagnose the problem and provide the student with an explanation of how to successfully solve problems and avoid making the identified mistake. Knowing the math at a "sophisticated level" gives the teacher a greater ability to do this.

Just for Fun. Why can't you trust a math teacher? They're always calculating.

CHAPTER 2. IMPORTANT STUFF: NUMBERS AND OPERATIONS

Building Over Time

Instruction and practice help students develop math skills. This development, or skill building, takes place gradually over time. Often, new skills build on previously learned skills. In this section, a Numbers and Operations Table is presented to serve as a guide for teaching skills and concepts at each grade. This guide may work fine as is for some teachers, whereas others may choose to shift some things around.

The table is not perfect or ideal, but if students are proficient with everything in the table, they should be well prepared for math in seventh grade. In comparing the content columns for each grade, sixth grade may appear to be light on content. There are a few things to keep in mind. Some schools/districts may be fairly rigid about specific content to be taught at each grade level. As a result, some teachers may shift some of the topics in the chart around. Some may find it will work better to shift some fourth-grade material to fifth grade and some fifth-grade material to sixth grade. It's a rare year when students entering a given grade level are fully prepared and proficient with the content and skills a teacher would like for them to have. As a result, time may not only be spent on review but backfilling deficiencies. So, when considering the table, factor in the reality of the school setting, expectations at other grade levels, and content and skills of the students entering the classroom.

NUMBERS AND OPERATIONS TABLE

	Kindergarten	Grade 1	Grade 2	Grade 3	Grade 4	Grade 5	Grade 6
Numbers							
Write	0–20 Count 1–100 Count back from 10	0–100	Up to 1,000				
Read/write				To six digits	To nine digits	To billions	To trillions
Compare and order	Sort, classify, sets, patterns	<, >, and = with objects and numbers	<, >, and = to 1,000	To six digits	To nine digits	To billions	To trillions
Place value		Ones, tens, hundreds	To thousands Expanded form	To hundred thousands	To hundred millions and thousandths	To billions and thousandths	To hundred billions
Miscellaneous		Ordinal position Identify even and odd numbers	Identify even and odd numbers	Ordinal position Perfect squares to 100			
Round to nearest			10	100	1,000	100,000 and thousandths	Millions
Exponents					Perfect squares to 100	Squared, cubed, nth Powers of ten to 10^6 Notation	

	Kindergarten	Grade 1	Grade 2	Grade 3	Grade 4	Grade 5	Grade 6
Fractions	Halves	Part of whole Part of group 1/2, 1/3, 1/4	1/2, 1/3, 1/4, 1/5, 1/6, 1/8, 1/10	Numerator Denominator Mixed numbers Equivalent Compare with like denominators	1/12 Convert between improper fractions and mixed numbers Lowest terms—simplify/reduce Rename with common denominators Compare with like and unlike denominators Add and subtract with like and unlike denominators Solve in form of 2/3 = ?/12	Lowest Common Denominator (LCD) with unlike denominators Add and subtract mixed numbers and fractions with unlike denominators Multiply and divide fractions and mixed numbers Round to nearest whole	
Decimals				Equivalents for 1/4, 1/2, 3/4 Read/write to hundredths	Read/write decimals as fractions Decimal equivalent for 1/8, 1/10 Expanded form Compare Add/subtract with two-place decimals	Read/write decimals to thousandths Expanded form Estimate sums/difference Multiply decimals by 10, 100, 1,000, another decimal Divide decimals by whole numbers and decimals	

	Kindergarten	**Grade 1**	**Grade 2**	**Grade 3**	**Grade 4**	**Grade 5**	**Grade 6**
Percents						Equivalences between fractions, decimals, and percents 1/10, 1/4, 1/2, 3/4 Find percent of a number	Find percent one number is of another number Find unknown numbers
Ratio and proportion						Ratio meaning Three forms Equal ratios	Unit rates Map scales and scale drawings Solve proportions Cross-products
Primes, composites, factors, and multiples					Introduction to multiples and factors	Introduction to prime numbers, and composite numbers Greatest common factor Least common multiple	Prime and composite numbers under 100 Prime factors of numbers to 100 Exponential notation for multiple primes
Integers					Introduction to negative integers	Add/subtract negative integers compare	Sum of opposites is zero Multiply/divide positive and negative integers
Addition	To 10	Facts up to 20 Two-digit without regrouping	Two- and three-digit with regrouping Commutative property Identity property	With regrouping to 10,000 Associative property Estimate			

	Kindergarten	Grade 1	Grade 2	Grade 3	Grade 4	Grade 5	Grade 6
Subtraction	From 10	Facts within 20 Two-digit without regrouping Inverse of addition	Two- and three-digit with regrouping	With regrouping within 10,000 Check by adding Estimate			
Multiplication	Skip count by twos, fives, tens		Skip count by threes Introduction Factor and product Facts for 1, 2, 3, 5, 10 Commutative property Identity property Zero property	Skip count by fours Facts to 10 x 10 Three digits by one digit with regrouping Multiplication and division as inverse operations	Mentally by 10, 100, 1,000 Identify multiples and common multiples Multiply two- and three-digit numbers Associative property Estimate	Distributive property Estimate	
Division				Dividend Divisor Quotient Remainders Facts to 100/10 Division by zero is undefined Identity property Multiplication and division as inverse operations Divide two- and three-digit numbers by one-digit divisors Check by multiplying	Ways to write division problems Identify factors, common factors Divide up to four digits by one- and two-digit divisors Check by multiplying Estimate	Divide by 10, 100, and 1,000 Divide up to four digits by up to three-digit divisors	

Key Grade-Level Topics

Grade level by grade level, I indicate what I feel are the key numbers and operations topics. This does not downplay the importance of other topics that may be significant building blocks. The key topics here are ones that should be adequately addressed. Often, other topics support or help develop the key topics. It is not intended for this section to address all topics but just a few key topics for each grade level. Refer to the Numbers and Operations Table above by grade level for topics not addressed in the text of this section.

Kindergarten. It is important for students in kindergarten to have many opportunities to count, compare, combine objects, work with patterns, and add and subtract to and from 10. Those opportunities should include working with and relating numbers to manipulatives, concrete objects, and pictures. This helps students begin to develop a good number sense that will serve them well. Skip counting by twos, fives, and tens is an early start on building a foundation for multiplication, learning about multiples, and learning multiplication facts.

Grade 1. Instruction in first grade should help students further develop number sense with comparing and ordering numbers, learning place value to the hundreds, learning simple fractions, and understanding addition and subtraction facts up to and within 20.

Grade 2. Instruction in second grade should build on previous learning experience to further develop number sense. Instruction and learning experiences should help students expand knowledge and understanding of place value to 1,000; begin to round numbers, learn fractions $\frac{1}{2}, \frac{1}{3}, \frac{1}{4}, \frac{1}{5}, \frac{1}{6}, \frac{1}{8}$ and $\frac{1}{10}$; learn multiplication facts for 1, 2, 3, 5, and 10; and add and subtract two- and three-digit numbers with and without regrouping using the standard algorithm.

Grade 3. In third grade, instruction and learning experiences should help students further expand their knowledge and ability to work with place value, rounding, and fractions. Fraction work should include being able to recognize and use equivalent fractions, comparing and ordering fractions, and mixed numbers. Students should be able to add and subtract up to and within 10,000 with regrouping using the standard algorithm. Students should begin to be familiar with decimals. Ability with multiplication facts to 10 × 10 should be developed as well as multiplying three-digit numbers by one-digit with regrouping. Division is a key topic at this grade level. Students should develop knowledge and ability to divide two- and three-digit numbers by one- digit divisors using the standard algorithm.

Grade 4. In fourth grade, instruction and practice should help students further expand and develop knowledge and ability working with decimals and multiplication. Fractions and long division are key topics for this grade. A major focus should be on multiple-digit multiplication, multiple-digit division using the standard algorithm, converting and simplifying/reducing fractions, and adding and subtracting decimals and fractions with common and unlike denominators.

Grade 5. In fifth grade, key topics include fractions, decimals, division, and integers. Work with fractions should focus on finding a least common denominator (LCD), greatest common factor (GCF), adding and subtracting mixed numbers and fractions with unlike denominators, and multiplying and dividing fractions and mixed numbers. Work with decimals should expand to include estimating sums and differences, multiplying, and dividing. The ability to divide up to four digits by a three-digit divisor should be developed. Adding and subtracting negative and positive integers should also be included.

Grade 6. In sixth grade, key topics include percent, ratio and proportion, primes, composites, factors, multiples, and negative integers. These topics build upon and use knowledge and skills developed in previous grades. Work with integers should include multiplying and dividing positive and negative numbers. Time should also be spent reviewing, maintaining, and backfilling as necessary knowledge and skills from previous grades with lots of attention given to multiplication facts, long division, and fractions. Mastery of the four arithmetic operations with whole numbers should be reached. If time permits, students may be ready for some pre-algebra and might work with exponents and begin learning to solve one-step equations.

There are math topics that are important but not included as a part of numbers and operations. These topics are measurement (linear, weight, money), geometry, and data/probability. These are important topics—they are wheat, not chaff—and should be taught. Telling time is one topic that is not part of numbers and operations but will be addressed here.

Time Does Not Tell Itself

Telling time is an important topic that deserves some attention for good reason. Even in a digital age, telling time with analog watches and clocks still needs to be explicitly taught. Too many adults these days can't tell time via analog devices. Unfortunately, some widely adopted math programs do not include telling time, so it was never taught to students in the classroom.

I remember substitute teaching in a senior statistics class. Toward the end of the class period many students would quickly take their cell phones out, look at them, and put them away. They knew they weren't supposed to have their cell phones out during class time. I nicely reminded a few students of this, and they let me know they were just checking the time. I pointed out to them that there was a clock within easy view on the nearby wall. They told me they couldn't tell time with that clock. It was analog, and they could only tell time when presented digitally. Nobody had taught them analog time.

I know of university graduate students who could not tell time with analog clocks. When placed in hospital internships, they would be late for appointments. One student blamed his lack of punctuality on the hospital for having analog clocks, which obviously would not help students who had never learned to read an analog clock.

There are commercially available teaching clocks teachers can use to demonstrate as well as classroom sets of student clocks. Some teachers like to ask students to make paper plate clocks to work with in class.

Here is a table that may serve as a guide for instruction on telling time.

	Kindergarten	Grade 1	Grade 2	Grade 3	Grade 4
Telling Time on Analog Clock	Recognize numbers on the clock and tell time to the hour	Tell and write time to the hour and half hour Elapsed time to the hour and half hour	Tell and write time to five minutes Use a.m. and p.m. Elapsed time to five minutes	Tell and write time to one minute Tell time to nearest half and quarter hour Add and subtract time down to the minute	Measure time in seconds

Some teachers may want to include adding and subtracting time. Adding and subtracting time can provide an opportunity to help students better understand place value and borrowing/carrying or regrouping. Here are a couple sample problems.

22 hr 18 min	10 hr 09 min
<u>+18 hr 29 min</u>	<u>–05 hr 24 min</u>

Just for Fun. What do you get if you cross a math teacher and a clock? Arithma-ticks!

A mathematician wandered home at 3 a.m. Her husband became very upset, telling her, "You're late! You said you'd be home by 11:45! The mathematician replied, "I'm right on time. I said I'd be home by a quarter of 12."

CHAPTER 3. NUMBERS YOU CAN COUNT ON

Numbers, Numerals, and Digits

A number is a count and is really a concept or idea in our minds. A numeral is a symbol used to write down or represent the conceptualized number. Try to get primary students to wrap their heads around that. Although it may be important to let students know the difference, simplifying may help. A number is in your head, and a numeral is what you write down. Numerals are made from symbols, and a single symbol is called a digit. In our base 10 system, there are 10 digits: 0, 1, 2, 3, 4, 5, 6, 7, 8, 9. A distinction is seldom made between a number and a numeral in common use today. In teaching students and in this text, I fairly consistently use the term "number" rather than "numeral."

Terminology and Expectations

I always let my students know the terminology I will use, especially if it differs from more technically correct terminology. Sometimes I won't need to let them know, such as when, just for fun, I will use carburetor for numerator and terminator for denominator. I will be clear about my expectations for how students should present their work, that is, format for working a problem, what and how work should be shown, and how to write numbers. For an example of writing numbers, I let students know that eight tenths might be written as .8 or 0.8 and that I accept it either way. I tell them they may have a teacher down the line who might want it a specific way and to do it the way their teacher expects it done.

Dieting Fractions

Reduce or simplify fractions? Either term works for me, and I feel students should be comfortable with the terms being used interchangeably. Although it is taught as an isolated skill, reducing fractions is a valuable assist when working with more complex problems that are easier to solve if smaller numbers are used. If I solve a problem requiring me to add, subtract, multiply, or divide fractions, I would rather deal with $\frac{2}{3}$ than, say, $\frac{488}{732}$. If I don't reduce $\frac{488}{732}$, I have some ugly work to do. I prefer to ask students to reduce to simplest terms. Put that way, reduce is a part of simplify. When I use the term "reduce," I often tell the students I'm putting the fraction on a diet. Let's get those fractions slimmed down!

Counting, Reading, and Writing Numbers (Numerals If You Must)

Students need lots of experience counting. Just reciting numbers, such as in a sing-song manner, is great, but it is not counting. Reciting does not involve awareness of the one-to-one correspondence that takes place when counting how many items are in a group of things. In second grade, students should be able to count up to 1,000. As they learn place value, students should be able to count even higher. Even though the expectation is there, I don't recall ever asking a student to count from 1 to 1,000.

Reading and writing numbers seem to follow counting. Students can be given opportunities to count a group of objects, write the number down, and read the number.

Even though number lines aren't mentioned or emphasized in addressing various topics here, they can play an important role in helping students. They can help students see numbers relative to one another and help compare and order numbers.

Compare and Order

Students benefit from lots of practice comparing and ordering numbers. They should use the appropriate symbols, <, >, =, when comparing numbers. I would begin practice comparing one-digit numbers, then move on to two-digit numbers, then three-digit numbers, and more as appropriate for the students. Place value knowledge comes in handy for students when they begin comparing numbers greater than single digits. For the < and > symbols, I would draw an alligator with the symbol as the mouth and tell students the alligator always is turned with its mouth open to the greatest number of chickens.

If students are able to compare numbers, given a set of numbers they should be able to order them by arranging them from least to greatest or from greatest to least. I have always enjoyed ordering numbers: "I'd like a 56 and a 72 with fries and ketchup on the side."

Place Value

The only time I had students really experience difficulty with place value was when I had to teach second-graders the base eight system. I think the intention was that it would help students better understand place value. The only thing it did a good job of was confusing students.

Students do need to know and understand place value in our base 10 system. They begin learning this by counting, skip counting by 10 and 100, and writing numbers. In primary grades, lots of experience should be provided in grouping objects in sets of tens and then hundreds, thousands, and so on. I made bean sticks to help students with this and then later, when available, used Cuisenaire rods and base 10 blocks. Making exchanges of 10 ones for one 10 not only helps students with place value, but also helps students when they begin addition and subtraction that requires regrouping. Relating how to represent groupings with place value is important.

Students benefit from practice converting between standard form and expanded form of numbers.

Standard Form Expanded Form

$$42{,}354 = 40{,}000 + 2{,}000 + 300 + 50 + 4$$

I start with two-digit numbers and gradually work up to three-, four-, and five-digit numbers.

$16 = 10 + 6$

$824 = 800 + 20 + 4$

$4{,}980 = 4{,}000 + 900 + 80 + 0$

$87{,}301 = 80{,}000 + 7{,}000 + 300 + 0 + 1$

It is important that students recognize that zeros may be place holders and can make a difference. Without zeros holding the place, 80,003 becomes 83. There is a difference between the 8 representing 80,000 and the 8 representing 80.

Below is an example of a number with each place value position labelled. The same number is written out in expanded form and added. Although this goes up to the billions, I would start with ones and tens and gradually work up to larger numbers (see Figures 3-1 and 3-2).

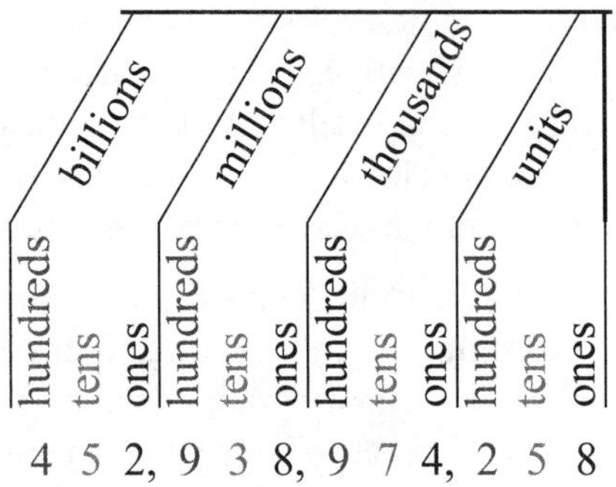

Figure 3-1

400,000,000,000	four hundred billion
50,000,000,000	fifty billion
2,000,000,000	two billion
900,000,000	nine hundred million
30,000,000	thirty million
8,000,000	eight million
900,000	nine hundred thousand
70,000	seventy thousand
4,000	four thousand
200	two hundred
50	fifty
+ 8	eight
452,938,974,258	

Figure 3-2

In working examples of the four basic operations, whenever possible, I point out how I write the numbers so the place values are in line.

During scratch-paper time, I would write a number such as 824,592 down for the students. Then, I might ask them to write down the digit in a given place, say the thousands place. Or, I might ask them to write down the number represented by the digit in the ten thousands place. I might read off a number and ask students to write the number down. In reading off a number such as "eighty thousand, twenty-three," I would caution students not to let me trick them. Seems like there is value in zero.

Place value for decimals is addressed in the decimal section.

Just for Fun. Figure out what my number is. The hundreds digit is an even number that's > 5 and < 9. It has 2 tens and 4 ones. It is not 624. What is my number?

Rounding

I start off simple with rounding one-digit numbers and then gradually moving to two digits, three digits, and so on. Place value knowledge is key, especially since the place value to round to is usually designated. As an example, I might ask students to round to the nearest 10, or the nearest 100,000.

The steps to rounding are 1) identity the digit in the place value position being rounded to, 2) look at the digit to the immediate right of that digit, 3) if it is less than 5, replace it and all digits to its right with zeros, and 4) if it is 5 or greater, bump the digit in the place value position being rounded up by 1, with all digits to the right becoming zeros.

Example 1. Round 4,592 to the nearest 100.

Step 1. Five is the digit in the hundreds place.

Step 2. Nine is the digit to the immediate right of the 5.

Step 3. Nine is not less than 5, so go to Step 4.

Step 4. Nine is 5 or greater, so bump the 5 in the thousands place up by 1. This makes it become a 6. All of the digits to the right of the 6 become zeros.

4,592 rounded to the nearest 100 is 4,600.

Example 2. Round 4,258 to the nearest 1,000.

Step 1. Four is the digit in the thousands place.

Step 2. Two is the digit to the immediate right of the 4.

Step 3. 2 Two is less than 5, so it and all the digits to the right become zeros while the 4 stays the same.

Step 4. Two is not 5 or greater, so stop at Step 3.

4,258 rounded to the nearest 1,000 is 4,000.

I used to have students who had difficulty rounding. They could find the digit in the place value that they needed to round to, but they had difficulty determining what to do next. I realized I needed to do a better job of teaching this skill. Rather than having the lesson be cut and dry, I started spicing it up by verbally animating the digits. In other words, I would have the numbers talk to each other with different voices that I would provide.

Here's an example of rounding 4,254 to the nearest 1,000 using talking numbers. The 4 says, "I've got to figure out whether to stay the same or to grow up and become a 5." So, the 4 looks next door to their neighbor on the right and sees a 2. The 4 shouts to get the 2's attention and says, "Hey, 2, are you less than 5 or 5 or greater? If you are less than 5, I will stay a 4. If you are 5 or greater, I will become a 5." The 2 responds, "Silly 4, can't you tell I'm less than 5?" The 4 says, "Thanks. You are a great help. Since you are less than 5, I will stay a 4." The 2 says, "Okay, so what does that mean for me?" The 4 answers, "Oh, I forgot to tell you. You will become a 0. I know you want to be well rounded. You will have some company because the digits behind you will also become zeros."

Rounding is used when estimating. I would encourage my students to try the Grocery Store Challenge. I would challenge them to use their rounding skills when they go to the store with a parent or guardian. They were to round the cost of each item to the nearest $1 or $10 and keep a running total in their head. They should compare it to the actual total at the checkout stand.

Just for Fun. A talking sheepdog rounds up all the sheep into the pen for his farmer. He reports back to the farmer, "All 40 sheep are accounted for." The farmer says, "But I've counted them, and there are only 36!" The sheepdog replies, "I know, but I rounded them up."

Order of Operations

Barry covers the order of operations well in Grade 7 (see Part II, Chapter 6). I will briefly cover some of the same information he provides as well as add a bit from my perspective and experience.

I always taught the order of operations to sixth-grade classes using PEMDAS. When I would initially teach it, it was before students had learned about exponents. I would just let them know they would learn about exponents later in the year.

P	**P**arentheses	First, do all operations within parentheses.
E	**E**xponents	Do all work with exponents.
M	**M**ultiply	Multiply and divide in order from left to right.
D	**D**ivide	
A	**A**dd	Add and subtract in order from left to right.
S	**S**ubtract	

I would stress the importance of always going through problems from left to right step by step. To complete the steps, students would go through each problem four times.

Step 1. **P**arentheses. Look through the problem from left to right and do all the operations within parentheses first. I would let my students know the term "parentheses" used here means any grouped or enclosed operation. If a grouping is nested within another, the inner-most grouping gets worked first.

Example.

$4 - 3[4 - 2(6 - 3)] \div 2$	$(6 - 3)$ This grouping is inner-most, so it gets worked first.
$4 - 3[4 - 2(3)] \div 2$	$[4 - 2(3)]$ This grouping is within the other grouping, so do the multiplication first.
$4 - 3[4 - 6] \div 2$	$[4 - 6]$ This completes the work inside groupings.
$4 - 3[-2] \div 2$	

Step 2. **E**xponents. Look through the problem from left to right and do all work involving exponents in the order they come when going from left to right.

$4 - 3[-2] \div 2$	There are no exponents, so nothing needs be done in this step.

Step 3. **M**ultiply and **D**ivide. Look through the problem from left to right and multiply and divide in order of appearance. This step needs additional attention to help students realize that if they come to a division operation first, they need to do it before moving on. Initially, some students will think they need to do multiplication first and then do division, so it is important they understand to do the operations in order of appearance. I might say, "Multiply or divide, whichever comes first."

$4 - 3[-2] \div 2$	$3[-2]$ This multiplication appears before the division.
$4 + 6 \div 2$	$6 \div 2$ This division comes next.
$4 + 3$	

Step 4. **A**dd and **S**ubtract. Look through the problem from left to right and add and subtract in order of appearance. This step may need additional attention to help students realize they need to do whichever operation comes first in the problem. I might say, "Add or subtract, whichever comes first."

$4 + 3$	The addition needs to be done; there is no subtraction to do.
7	

I start with simple problems and then work up to more advanced ones, such as the above example. The example calls for students to be able to multiply negative integers, which most sixth-graders have not worked with when they begin learning the order of operations.

To emphasize the point of doing whichever comes first in Step 3 and Step 4, I work through examples where division comes before multiplication and subtraction comes before addition.

I like how Barry would ask students to place parentheses in expressions so that the expressions follow the order of operation rules. I wish I had known to do that. I did ask students to do something similar by giving them equations and asking them to place parentheses as needed to make the equations true.

Example

$14 \div 2 + 5 - 1 = 1$

PEMDAS worked well for me and my students. I knew other teachers who used BODMAS for Brackets, Orders, Division, Multiplication, Addition and Subtraction.

I let students know they could remember PEMDAS by saying "**P**lease **E**xcuse **M**y **D**ear **A**unt **S**ally." No student ever asked what Aunt Sally had done or was going to do that she needed to be excused for. It's the kind of thing I would ask about.

Just for Fun. A surgeon says to the nurse, "I have so many patients. Who do I work on first?" The nurse responds, "Simple, follow the order of operations."

CHAPTER 4. MATH PROPERTIES: NUMERICAL REAL ESTATE?

Students need a working knowledge of math properties. By working knowledge, I mean they should understand and be able to apply the properties. Although I use the property names and expect students to be familiar with the names, it is enough for me if students can successfully use the properties and say what the property is or does even if they can't name it.

In this section, I will present the properties elementary students need to learn. I will present the property name, explain what it means, provide an example using numbers, and provide a way that may help students remember some property names.

Identity Properties

A number maintains its identity, or stays the same, upon the completion of an operation using a number that is referred to as the "identity" of the completed operation. With this property, there is no identity theft.

Identity Property of Addition. Zero is the additive identity. Any number plus 0 is equal to the number 0 is added to.

$5 + 0 = 5$ $\qquad$ $847{,}208{,}824 + 0 = 847{,}208{,}824$

Identity Property of Subtraction. Zero is the subtraction identity. Any number minus 0 is equal to the number 0 is being subtracted from.

$5 - 0 = 5$ $\qquad$ $847{,}208{,}824 - 0 = 847{,}208{,}824$

Identity Property of Multiplication. One is the multiplication identity. Any number multiplied by 1 is equal to the number being multiplied by 1.

$5 \times 1 = 5$ $847{,}208{,}824 \times 1 = 847{,}208{,}824$

Identity Property of Division. One is the division identity. Any number divided by 1 is equal to the number being divided by 1.

$5 \div 1 = 5$ $847{,}208{,}824 \div 1 = 847{,}208{,}824$

Commutative Property

In the commutative property, the order of numbers in the operation does not matter. The position of the numbers in the operation can be changed without affecting the answer. The commutative property applies to addition and multiplication. It does not apply to subtraction and division. The word itself can help students remember the property. To "commute" is to travel or move from one location to another. In this case, the numbers travel or move around without affecting the answer.

Commutative Property of Addition. The order of the numbers in addition does not matter. They can be moved around.

$$5 + 3 = 3 + 5$$

$5 + 3 = 8$ $3 + 5 = 8$

$2 + 6 + 8 + 4 = 8 + 2 + 6 + 4$ $2 + 6 + 8 + 4 = 20$ $8 + 2 + 6 + 4 = 20$

Commutative Property of Multiplication. As with addition, the order of the numbers in multiplication does not matter. They can be moved around.

$3 \times 4 = 12$ $4 \times 3 = 12$

$5 \times 6 \times 2 = 2 \times 5 \times 6$ $5 \times 6 \times 2 = 60$ $2 \times 5 \times 6 = 60$

The commutative property does not apply to subtraction.

$$5 - 3 \neq 3 - 5$$

$5 - 3 = 8$ $3 - 5 \neq 8$

The commutative property does not apply to division.

$$8 \div 2 \neq 2 \div 8$$

$$8 \div 2 = 4 \qquad 2 \div 8 \neq 4$$

Associative Property

The associative property applies to addition and multiplication when there are three or more numbers involved in the operation. Unlike the commutative property, where numbers change location, in the associative property the numbers are grouped. In the associative property, the results will be the same even if the grouping changes. Let's consider an example using friends who associate with each other. I'll use three boys. Joe, Sam, and Rick are going to end up together but can get together in two different ways.

(Joe + Sam) + Rick. In this case, Joe and Sam are together and associating with each other already, and then they are joined by Rick.

Joe + (Sam + Rick). Here, Sam and Rick are together and associating with each other already, and then Joe joins them.

Associative Property of Addition. When three or more numbers are added, numbers can be grouped without affecting the result.

$$5 + 3 + 7 = 15$$

$$(5 + 3) + 7 \qquad 5 + (3 + 7)$$

$$8 + 7 = 15 \qquad 5 + 10 = 15$$

$$5 + 3 + 7 = (5 + 3) + 7 = 5 + (3 + 7)$$

Associative Property of Multiplication. When three or more numbers are multiplied, numbers can be grouped without affecting the result.

$$5 \times 3 \times 2 = 30$$

$$(5 \times 3) \times 2 \qquad 5 \times (3 \times 2)$$

$$15 \times 2 = 30 \qquad 5 \times 6 = 30$$

$$5 \times 3 \times 2 = (5 \times 3) \times 2 = 5 \times (3 \times 2)$$

Zero Property

The zero property is a property unique to the operation of multiplication.

Zero Property of Multiplication. Any number multiplied by 0 is 0.

$5 \times 0 = 0$ $0 \times 824 = 0$ $4785 \times 0 = 0$

Distributive Property

The distributive property can be used to solve problems such as 3(4 + 1). Although this is a simple problem to work without using the distributive property, not all problems are as simple. In this problem, the three would be distributed to each term in the parenthesis.

$3(4 + 1) = 3 \times 5 = 15$

$(3 \times 4) + (3 x 1) = 12 + 3 = 15$

$3(4 + 1) = (3 \times 4) + (3 \times 1)$

As repeated addition, $3(4 + 1) = (4 + 1) + (4 + 1) + (4 + 1) = 5 + 5 + 5 = 15$.

Here's an example that shows how students might use their place value knowledge in working a problem using the distributive property.

The expanded form of 824 is 800 + 20 + 4. Using the distributive property, 2 × 824 can be rewritten as 2(800 + 20 + 4).

Using the distributive property, 2 × 824 can be calculated this way:

$2(800 + 20 + 4) = (2 x 800) + (2 x 20) + (2 x 4) = 1600 + 40 + 8 = 1648$

So, 2 × 824 = 1648.

Just for Fun. Why do subtraction and division work from home? They don't commute.

CHAPTER 5. ZEROS: HERE GOES NOTHING

The Value of Zero

In this section, I want to provide information about zeros and how place value knowledge related to zeros helps students solve many problems easily and quickly. It is important that students understand place value and why these strategies work. What, to some, may appear to be tricks or gimmicks are really strategies based on place value. It is acceptable for students to make good use of these strategies when they understand why they work.

Suggested grade levels for teaching these strategies aren't necessarily provided here. I suggest you use your own judgment as to when to include instruction about these strategies. The sequence chart can provide some guidance as to when information presented in this section is appropriate. Some textbook series may address zeros well and provide logically sequenced practice, whereas other series may address very little of what is presented here.

Zeros or Zeroes?

This is one of those words that appears to be correct either way in most cases. If used as a verb, it should be spelled "zeroes." In this section about zeros, it isn't being used as a verb. I like potatoes and tomatoes, so my tendency would be to use "zeroes." I also like tacos, so I'm comfortable using zeros. A mathematician I respect uses zeros, so I have chosen to use "zeros" for the plural spelling. If you have five zeros, do you have 5 or 0?

ZEROS AS PLACE HOLDERS, COMPARISON HELPERS, AND TRAILING ZEROS

Students should be able to recognize when zeros serve as place holders. As an example, 92 is not the same as 902 or 920. In 902, the 0 is holding the tens place, for which there are 0 tens. In 920, the 0 is holding the ones place, for which there are 0 ones.

Students may find it helpful to add one or more zeros in front of a number when comparing it to another number. When comparing numbers to see which is greater or lesser, students need to first identify the largest place value represented in the numbers being compared. As an example, in comparing 2,454 and 987, students should identify the thousands place as the highest place value in 2,454. Since 987 doesn't have a digit in the thousands place, students can put a zero there to help with the comparison. It should now be easier to compare 2,454 and 0,987 by looking at the digits in the thousands place: 2 > 0, so 2,454 > 0,987 (or 987 if the 0 is dropped). I stress the importance of using the digits in the same place value. Some students tend to compare the first digits of each number without regard for place value. Looking at it this way, some students will see 2 < 9 and then say 2,454 < 987; 2,454 is greater than, not less than, 987.

Zeros appearing at the end of a number are referred to as trailing zeros. The zeros that appear in these numbers are trailing zeros: 30; 2,840,000; 8,700. The zero in 507 is a place holder and not a trailing zero. In 340,772,000, the three zeros at the end are trailing zeros, whereas the zero between the 4 and 7 is a place holder.

ADDITION WITH ZERO

If students add any number to zero, they should get the number they started with. Students need to know that the sum of any whole number added to zero will be that whole number. This is called the additive identity. This holds true whether adding zero to a positive number or a negative number. Examples: 0 + 5 = 5. 0 + –5 = –5. Of course, the order in addition doesn't matter: 5 + 0 = 5. –5 + 0 = –5.

SUBTRACTION

If students take zero away from any number, their answer should be that number. Examples: $5 - 0 = 5$. $452 - 0 = 452$. Unlike addition, the order does matter. $0 - 5 \neq 5$ but equals -5. This is better addressed with students in instruction related to integers. The important thing here is for students to understand that taking 0 away from a number leaves them with the number they started with.

MULTIPLICATION

In this section, I'll first address what students need to know about the zero property of multiplication and then how students can multiply one-digit whole numbers by multiples of 10 and then build on that ability to be able to multiply multi-digit whole numbers by multiples of 10. At this point, understanding and ability is being developed to work with whole numbers since students haven't yet learned or worked with decimals. Multiplying with decimal numbers is addressed in Chapter 8.

Zero Property of Multiplication. Students need to know the product of any number multiplied by 0 is 0. This is the zero property of multiplication. Examples: $5 \times 0 = 0$, $0 \times 824 = 0$. When I address this, I take the opportunity to emphasize the importance of looking over the whole problem before beginning any computation. I encourage students to look for easy ways to solve a problem. Why do a bunch of work when you don't have to? Often, to start scratch-paper time I might present a problem like this: $24 \times 4 \div 7 \times 2 \times 0 \times 25 \div 5$. Most students will see the answer is zero. Incorporating learning that has already taken place, I might construct more complex problems using fractions and decimals. My students seemed to enjoy these kinds of problems. I think it helped them look for easy ways to tackle what initially appeared impossible and made them more comfortable when setting out to work larger, seemingly ugly, problems.

Trailing Zeros: Multiplying One-Digit Numbers by Multiples of 10. Students can begin work on multiplying one-digit numbers by multiples of 10 as they learn their multiplication facts. They should already be able to count by tens to 100 and possibly beyond.

$1 \times 10 = 10$	$6 \times 10 = 60$
$2 \times 10 = 20$	$7 \times 10 = 70$
$3 \times 10 = 30$	$8 \times 10 = 80$
$4 \times 10 = 40$	$9 \times 10 = 90$
$5 \times 10 = 50$	$10 \times 10 = 100$

Students should be able to work any of these problems as repeated addition.

Example: $7 \times 10 = 10 + 10 + 10 + 10 + 10 + 10 + 10$. This problem can also be shown as $10 \times 7 = 7 + 7 + 7 + 7 + 7 + 7 + 7 + 7 + 7 + 7$. Both problems can be worked out, and both yield 70 as an answer. I would ask students which repeated addition problem they would rather work and why. I would anticipate they would rather do the work adding tens.

I ask students lots of questions about what they know (or think they know) and are observing. What happens when you multiply a number by 10? Does the identity property of multiplication apply to multiplying by 10? Students should begin to realize that multiplying any number by 10 can be done by adding 0 to the end of the number. I build on this to help students multiply multi-digit numbers times multiples of 10 and then to multiply multiples of 10 times multiples of 10.

Trailing Zeros: Multiplying Multi-digit Numbers by Multiples of 10. The previous section dealt with multiplying a one-digit number times a multiple of 10. I ask students to build on that to multiply multi-digit numbers times multiples of 10.

I may show a buildup and ask students to work on some problems like these.

$57 \times 10 = 570$	$23 \times 10 = 230$	$8245 \times 10 = 82450$
$57 \times 100 = 5700$	$23 \times 100 = 2300$	$8245 \times 100 = 824500$
$57 \times 1000 = 57000$	$23 \times 1000 = 23000$	$8245 \times 1000 = 8245000$
$57 \times 10000 = 570000$	$23 \times 10000 = 230000$	$8245 \times 10000 = 82450000$

I might ask students questions about each problem. How many trailing zeros are in the problem? In the answer? Do you get the same result if you do repeated addition?

Trailing Zeros: Multiplying Multiples of 10 by Multiples of 10. I ask students to take advantage of knowledge they have learned. I ask how many trailing zeros are in 10 × 10. I build on this to have students work up to problems such as 570 × 10 = 5700, 230 × 10 = 2300, 82450 × 10 = 824500. Then, I ask them to start tackling problems with more trailing zeros.

57 × 10 = 570	23 ×10 = 230	8245 × 10 = 82450
570 × 100 = 57000	230 ×100 = 23000	82450 × 100 = 8245000
5700 × 1000 = 5700000	2300 × 1000 = 2300000	824500 × 1000 = 824500000
57000 × 10000 = 570000000	23000 × 10000 = 230000000	8245000 × 10000 = 82450000000

I ask students to look at each problem above and answer questions about each one. How many trailing zeros are in the problem? In the answer? Do you get the same result if you do repeated addition?

I ask students to look at a similar set a little closer. I start with 32, multiply it by 10, then multiply the result by 10, and continue until I have multiplied by 10 four times.

32 × 10 = 320

320 ×10 = 3200

3200 × 10 = 32000

32000 × 10 = 320000

I might ask students questions about what they see happening. I hope to get a response helping everyone see that each time I multiplied by 10 a trailing zero was added. Multiplying by 10 four times added four zeros, so I ended up with 320,000.

Using what students have learned from problems such as the above, I start presenting them with problems such as 800 × 400. My hope is that knowledge and understanding from previous work will help students bypass some work if they see this problem can easily be done by multiplying 8 × 4 and then adding four trailing zeros. I use scratch-paper time to work through several examples with an explanation of each step to help students better understand. As students become successful working things out step by step, I show and explain how some steps can gradually be dropped. In doing so, students may be able to solve problems such as 800×400 in their head.

By now, students should be comfortable with multiplying the non-zero digits and then adding the number of trailing digits found in the problem. Although it may be tedious, it doesn't hurt for me to show one or more problems worked out as repeated addition. Doing so may accomplish two things: 1) It will help reinforce student understanding of how and why this works, and 2) it will convince students to solve these kinds of problems using a strategy based on place value rather than working them out using repeated addition.

At this point, if students are performing successfully during scratch-paper time and on related assigned problem sets, I move the class on to division. If I feel more time needs to be spent on multiplication, however, I present a few more sets of problems such as those below.

$2 \times 3 = 6$	$25 \times 3 = 75$	$32 \times 3 = 96$
$20 \times 30 = 600$	$250 \times 30 = 7500$	$320 \times 30 = 9600$
$200 \times 300 = 60000$	$2500 \times 300 = 750000$	$3200 \times 300 = 960000$

I might vary these problems or ones like them to include multipliers with different numbers of trailing zeros. I might also show how problems such as these can be solved using repeated addition or by expanding them using knowledge of place value. Students have already seen and solved some problems using repeated addition, so they should know how that works. I show students how to solve $250 \times 30 = 7500$ by expanding:

$250 \times 30 = 7500$

$250 = 25 \times 10$ and $30 = 3 \times 10$

Therefore, 250×30 can be rewritten as $25 \times 10 \times 3 \times 10$.

Using the commutative property, $25 \times 10 \times 3 \times 10$ can be written as $25 \times 3 \times 10 \times 10$.

$25 \times 3 = 75$, which gives us $75 \times 10 \times 10$.

From previous work, students know that 75 × 10 will equal 750, giving us 750 × 10 = 7500.

$250 \times 30 = 25 \times 10 \times 3 \times 10 = 25 \times 3 \times 10 \times 10 = 75 \times 10 \times 10 = 750 \times 10 = 7500.$

As sets of problems are shown and worked on by me and the students, I ask students about what they see in the sets of problems. I ask if they see a quick way to solve problems based on what they understand. I do so in the hope such questioning will lead some to respond in a way that will lead others to see (if they haven't already) that you can simply multiply the non-zero digits and then add the number of trailing zeros to the product. It should be clarified that non-zero digits as used here should really read non-trailing zero digits, otherwise 20300 × 20 may be solved incorrectly as 23 × 2 × 100 × 10.

I feel it is important to show and explain to students how multiplying numbers with trailing zeros works and to help them understand why it does. Is it really a trick or gimmick if students employ sound understanding of mathematics to make their work easier (and more fun)?

Noggin Challenges

My students enjoyed challenge problems that I gave them. I sometimes asked them to work through problems and then would throw in a twist in the next problem, telling them not to let me trick them. If I told them that, they quickly learned to look for some little nuance that might trip them up. If they were tricked once, they were careful to not let me trip them up on the next problem. Not only did I do this with challenge-type problems, but also as a way to introduce the next step to our current work. At any rate, here are a few challenge problems that I used related to the zero property of multiplication. Paper and pencil were not required for problems such as these. I tell students this can be upstairs work—in other words, they can do these in their heads if they like.

Students have seen this one already.

1. $24 \times 4 \div 7 \times 2 \times 0 \times 25 \div 5.$

Try these:

2. $\frac{3}{4} \times 72 \div 4598.35667 \times 3 \times 0 \times 750 \div 3.3.$

3. (4/5 × 86 ÷ ½ × 0 x 453) + (824 ÷ 7 × 0 × 36900 × 43005).

Now, don't let me trick you.

4. ($\frac{4}{5}$ × 86 ÷ $\frac{1}{2}$ × 0 × 453 + 1) + (824 ÷ 7 × 0 × 36900 × 43005).

Since I provide answers in class so students can check how they did, it is only fair to do so here:

1. 0.
2. 0.
3. 0.
4. 1 (0 + 1 = 1).

I construct challenge problems related to our current work. I also ask students to construct some challenge problems.

Division

Students need to be able to work with division in fraction form and long division form. In this section, I ask students to look at and work with division in fraction form. I start off comparing fractions and division presented in fraction form.

$\frac{numerator}{denominator}$ corresponds to $\frac{dividend}{divisor}$, and $\frac{dividend}{divisor} = quotient$

Students can check their division work by repeated subtraction. Although I may ask students to check a few problems with this method, I move on to having them check their work by multiplying the divisor and the quotient. Once they have worked through checking some problems both ways, I ask questions about which method they prefer and why (i.e., which method is easier and quicker to use when working with larger numbers?).

$\frac{8}{2}$=?

8 – 2 = 6

6 – 2 = 4

4 – 2 = 2

2 – 2 = 0

Four twos were subtracted from 8. So, $\frac{8}{2} = 4$.

Students can also check their division work by multiplying the divisor and the quotient, which should equal the dividend if correct.

$$\textit{divisor} \times \textit{quotient} = \textit{dividend}$$

$\frac{8}{2} = 4 \qquad 2 \times 4 = 8$ correct.

I ask students to look at division with zero as the divisor and then division when the dividend is zero. I might present these two cases in words and then move on to considering them with numbers.

$$\frac{\textit{dividend}}{0} = \textit{quotient}; \qquad \frac{0}{\textit{divisor}} = \textit{quotient}$$

I explain to students that dividing by zero doesn't work and that it is undefined. I explain that 0 divided by any number is 0. Although I tell students this, many of them have a hard time wrapping their heads around it. For me, it was important to explain the concept and work examples together to help students understand. Here are some possible ways I use to explain the concept and show why it works.

Division by Zero: Mission Impossible. I remind students that multiplication and division are inverse operations. I also refresh their knowledge of the zero property of multiplication, which says the product of any number multiplied by 0 is 0. I explain that division does not have a zero property, like multiplication; if it did, it would read as the quotient of any number divided by 0 is 0. That would look like this:

$\frac{\textit{dividend}}{0} = 0$; Using 8 as the dividend $\frac{8}{0} = 0$

I show this won't work by multiplying the divisor, zero, and quotient, zero, to check the answer: $0 \times 0 = 0$. If the quotient were 0, that would mean that 8 should be the product of multiplying the divisor, 0, and quotient, 0. That doesn't happen since $0 \times 0 = 0$. Another way to look at it is to ask what number *can* we multiply by the divisor to get the dividend? In this case, any number multiplied by zero is zero; therefore, we will never get the dividend. So, we say division by zero is undefined.

I show students an attempt to check the work with repeated addition. Prior learning indicates that if 0 is taken away from any number, the answer will be that number. That is, 8 – 0 = 8. I show that I can subtract 0 from 8 an indefinite number of times and always get 8 as an answer. Zero will not work as a divisor. I relate the format being used here for division to fractions. I explain that if zero is the denominator, the fraction is undefined.

Zero as a Dividend: Nothing to It. Now, I show students what happens when zero is the dividend or numerator.

$$\frac{0}{divisor} = quotient = \frac{0}{denominator}$$

Using 9 as our divisor. $\frac{0}{9}$ = *quotient.* The result will be $\frac{0}{9} = 0$.

Dividing zero by any number (other than zero) is zero. I check students' work to help them understand why it is this way.

$divisor \times quotient = dividend: 9 \times 0 = 0$.

That works, so it is correct.

$\frac{0}{9} = 0$. This can be checked by multiplying both sides by 9.

$9 \times \frac{0}{9} = 0 \times 9$. This results in 0 = 0, which is true.

Checking by repeated subtraction may seem trickier.0 – 9 = –9. Nine can be subtracted from zero an indefinite number of times and will never yield zero as an answer. There is one, and only one, number that can be subtracted from zero that will yield zero, and that number is zero (i.e., 0 – 0 = 0).

I find it easier to just ask students to solve story problems such as this one: You have 0 pieces of candy that you divide evenly among your 9 friends. How many pieces of candy will each friend get?

Division with Trailing Zeros

This section will consist of two parts. The first will look at dividing multiples of 10 by one-digit numbers. The second will look at dividing multiples of 10 by multiples of 10. Examples used in this section will be ones resulting in whole number quotients. Trailing zero problems having a divisor greater than the dividend or resulting in a decimal answer are briefly addressed.

Trailing Zeros: Division of Multiples of 10 by One-Digit Numbers. I remind students of the identity property of multiplication and explain that division also has an identity property. Any number divided by 1 will give the same quotient as the number itself (see Figure 5-1).

$$1\overline{)1\ 5}\quad\text{quotient } 1\ 5 \qquad \text{In fraction form: } \tfrac{15}{1} = 15.$$

Figure 5-1

In addition to showing a few problems with multiples of 10 being divided by one-digit numbers, I would probably ask students to work on a few problems in at least the I-We part of I-We-You. My students often enjoyed being given problems bigger than these examples that they could easily and quickly solve. They might initially groan, but once they really looked at the problem, they would be delighted.

The problems here are presented in both fraction and long division form. I start off using 1 as the divisor (see Figure 5-2).

$\tfrac{150}{1} = 150$ $\tfrac{1500}{1} = 1500$ $\tfrac{15000}{1} = 15000$ $\tfrac{150000}{1} = 150000$

$1\overline{)1\ 5\ 0}$ = 1 5 0 $\quad 1\overline{)1\ 5\ 0\ 0}$ = 1 5 0 0 $\quad 1\overline{)1\ 5\ 0\ 0\ 0}$ = 1 5 0 0 0 $\quad 1\overline{)1\ 5\ 0\ 0\ 0\ 0}$ = 1 5 0 0 0 0

Figure 5-2

Now, I use 3 as the divisor (see Figure 5-3). $\tfrac{15}{3} = 5$.

$$3\overline{)1\ 5}\quad\text{quotient } 5$$

Figure 5-3

And the process is continued for dividends with trailing zeros. (See Figure 5-4)

$\frac{150}{3} = 50$; $\frac{1500}{3} = 500$; $\frac{15000}{3} = 5000$; $\frac{150000}{3} = 50000$

$$3\overline{)150}=50 \quad 3\overline{)1500}=500 \quad 3\overline{)15000}=5000 \quad 3\overline{)150000}=50000$$

Figure 5-4

I usually have a constant patter of explanation and questions related to what is being worked on. Here are some of the questions I would ask.

"What do you see in these sets of problems? Do you see a quick way to work the problems based on what you understand?"

Such questioning (usually) leads some students to respond in a way that leads others to see (if they haven't already) that they can simply divide the non-trailing zero digits by a one-digit divisor and then add the number of trailing zeros in the dividend to the quotient.

During scratch-paper time, I show and work through some explicit examples and ask students to solve a few problems.

Trailing Zeros: Division of Multiples of 10 by Multiples of 10. I then show 150 ÷ 10 can be solved as a repeated subtraction problem to demonstrate that students can get to 0 after 10 has been subtracted 15 times from 150. I sometimes ask students to work on a few problems using repeated subtraction. By the time students are learning about division by multiples of 10 they should be familiar with division as repeated subtraction (shown below), so this only needs to be a refresher.

150 – 10 = 140	100 – 10 = 90	50 – 10 = 40
140 – 10 = 130	90 – 10 = 80	40 – 10 = 30
130 – 10 = 120	80 – 10 = 70	30 – 10 = 20
120 – 10 = 110	70 – 10 = 60	20 – 10 = 10
110 – 10 = 100	60 – 10 = 50	10 – 10 = 0

Seeing that both the divisor and dividend (or numerator and denominator when expressed as a fraction) have a trailing zero in the one's place, students learn that both can be divided by 10. I work toward students understanding that the result is the same as canceling the zero in the one's place in the divisor and dividend.

$\frac{15\varnothing}{1\varnothing} = 15 \qquad \frac{150\varnothing}{1\varnothing} = 150 \qquad \frac{1500\varnothing}{1\varnothing} = 1500 \qquad \frac{15000\varnothing}{1\varnothing} = 15000$

Here are the same problems worked out in long division format (see Figure 5-5).

```
      1 5           1 5 0            1 5 0 0             1 5 0 0 0
1 0|1 5 0     1 0|1 5 0 0     1 0|1 5 0 0 0     1 0|1 5 0 0 0 0
  - 1 0         - 1 0           - 1 0             - 1 0
    5 0           5 0             5 0               5 0
  - 5 0         - 5 0           - 5 0             - 5 0
      0             0 0             0 0 0             0 0 0 0
```

Figure 5-5

Here are the same problems worked out again in long division format showing the cancelling of the 0 in the one's place in the divisor and dividend (see Figure 5-6).

```
     1 5            1 5 0             1 5 0 0             1 5 0 0 0
1 Ø|1 5 Ø     1 Ø|1 5 0 Ø     1 Ø|1 5 0 0 Ø     1 Ø|1 5 0 0 0 Ø
  - 1           - 1             - 1               - 1
    5             5               5                 5
  - 5           - 5             - 5               - 5
    0             0 0             0 0 0             0 0 0 0
```

Figure 5-6

Problems have been shown here with divisors of 1, 3, and 10. Before moving on to divisors with more than one trailing zero, let's look at a problem using 30 as the divisor.

In 150 ÷ 30, there's one trailing zero in the divisor and one in the dividend. In terms of fractions, a trailing zero is canceled in the numerator and denominator. I would emphasize to students that by doing the same cancellation "upstairs and downstairs," things are kept in balance.

$\frac{15\varnothing}{3\varnothing} = 5 \qquad \frac{150\varnothing}{3\varnothing} = 50 \qquad \frac{1500\varnothing}{3\varnothing} = 500 \qquad \frac{15000\varnothing}{3\varnothing} = 5000$

Long division format without cancelling is shown in Figure 5-7.

$$\begin{array}{r} 5 \\ 30\overline{)150} \\ -150 \\ \hline 0 \end{array} \quad \begin{array}{r} 50 \\ 30\overline{)1500} \\ -150 \\ \hline 00 \end{array} \quad \begin{array}{r} 500 \\ 30\overline{)15000} \\ -150 \\ \hline 000 \end{array} \quad \begin{array}{r} 5000 \\ 30\overline{)150000} \\ -150 \\ \hline 0000 \end{array}$$

Figure 5-7

Long division format with cancelling is shown in Figure 5-8.

$$\begin{array}{r} 5 \\ 3\not{0}\overline{)15\not{0}} \\ -15 \\ \hline 0 \end{array} \quad \begin{array}{r} 50 \\ 3\not{0}\overline{)150\not{0}} \\ -15 \\ \hline 00 \end{array} \quad \begin{array}{r} 500 \\ 3\not{0}\overline{)1500\not{0}} \\ -15 \\ \hline 000 \end{array} \quad \begin{array}{r} 5000 \\ 3\not{0}\overline{)15000\not{0}} \\ -15 \\ \hline 0000 \end{array}$$

Figure 5-8

I ask students what they see in these sets of problems; that is, do they see a quick way to solve the problems based on what they understand? I do this in the hope that such questioning leads some students to respond in a way that leads others to see (if they haven't already) that you can simply cancel the same number of trailing digits from the divisor and dividend and then perform the remaining division operation.

I ask students to do some scratch-paper work with a few problems with more than one trailing zero. Here are examples presented in fraction form and long division form showing the zeros cancelled out (see Figure 5-9).

$$\frac{450\not{0}\not{0}}{9\not{0}\not{0}} = 50 \qquad \frac{36\not{0}\not{0}\not{0}}{12\not{0}\not{0}\not{0}} = 3$$

$$\begin{array}{r} 50 \\ 9\not{0}\not{0}\overline{)450\not{0}\not{0}} \\ -45 \\ \hline 00 \end{array} \qquad \begin{array}{r} 3 \\ 12\not{0}\not{0}\not{0}\overline{)36\not{0}\not{0}\not{0}} \\ -36 \\ \hline 0 \end{array}$$

Figure 5-9

Trailing Zeros: Division When the Divisor Has More Trailing Zeros than the Dividend. The trailing zero problems up to this point have worked out nicely. But what if they don't work out nicely? Say, when the divisor has trailing zeros and the dividend doesn't or when the divisor has more trailing zeros than the dividend.

In the example of 15 ÷ 30, the divisor has one trailing zero, and the dividend doesn't have any. Thirty does not go into 1 or 15, but the problem can still be solved by adding a decimal to the dividend, followed by as many zeros as needed. In this problem, only one zero needs to be added after the decimal. A decimal also goes upstairs in the answer, and then the problem can be solved like any other problem without the decimal tripping things up (see Figure 5-10).

$$30\overline{)\,15} \qquad\qquad \begin{array}{r} .5 \\ 30\overline{)\,15.0} \\ -\underline{150} \\ 0 \end{array}$$

Figure 5-10

In the example of 150 ÷ 3000, the divisor has three trailing zeros, and the dividend has one. Three thousand is larger than 150, but the problem can still be solved by adding a decimal to the dividend followed by as many zeros as needed. In this problem, two zeros need to be added after the decimal, resulting in the same number of trailing zeros in the dividend and divisor. As in the previous example, a decimal also goes upstairs in the answer so the problem can be solved (see Figure 5-11).

$$3000\overline{)\,150} \qquad\qquad \begin{array}{r} .05 \\ 3000\overline{)\,150.00} \\ \underline{0000} \\ 15000 \end{array}$$

Figure 5-11

Zeros: Something More than Nothing. The examples worked on in this section about zeros did not have remainders, with the exception of problems whose divisor had more trailing zeros than the dividend. Remainders are addressed in Chapter 7, which discusses division in detail. Problems involving decimals as well as zero issues related to decimals are addressed in Chapter 8.

What I Expect Students to Know Related to Zeros. Provided adequate explanation and experience, students should know and be able to work with the following:

- Additive identity: Adding zero to a number results in the original number.
- Subtractive identify: Taking zero away from a number results in the original number.
- Zero multiplied by any number is zero.
- Add zeros to help do multiplication with trailing zeros
- Cancel to help do division with trailing zeros
- Zero divided by any number is zero.
- Division by zero is undefined.

I am more interested in students being able to successfully apply what they have learned than their ability to verbalize an explanation of understanding how it works. A verbalized explanation of understanding is icing on the cake. Students successfully applying their learning is the goal and is acceptable as an indicator of a working understanding.

Just for Fun. How do you solve any equation? Multiply both sides by 0.

What did the 0 say to the 8? “Nice belt!”

CHAPTER 6. ADDITION AND SUBTRACTION

Students in Grade 1 should learn their addition and subtraction facts up to and within 20 and be able to add two-digit numbers without regrouping. By the end of Grade 2, they should be able to add and subtract two- and three-digit numbers with regrouping. At the end of Grade 3, they should be able to add and subtract with regrouping up to and within 10,000.

I started teaching second grade in the late 1970s. As a matter of course, in the second half of the year, I taught students to add and subtract two- and three-digit numbers with regrouping using the standard algorithm. Students were able to learn this and be successful with it. Other second-grade teachers successfully taught their students how to add and subtract with regrouping. I never heard any teachers say it was too hard to teach or that their students couldn't learn it.

This was at a time when the terminology had changed from "carrying" in addition and "borrowing" in subtraction to "regrouping." For me, the process is the same whatever terminology is used. I use the terms "regrouping," "carrying," and "borrowing" with students, making sure they know and understand the terms. Parents were often confused with the term "regrouping." I would just tell them it is the same as carrying and borrowing, and they would be okay with that.

In the early grades, work with manipulative objects is important. That work can progress from manipulatives to visuals and then on to abstract work. Students should learn how to use a number line to add and subtract. Tally marks can also be helpful. Students should have experience exchanging 10 for 10 ones and 10 ones for 10. This experience should eventually extend to exchanging 100 for 10 tens and 10 tens for 100. Bean sticks, Cuisenaire rods, and base 10 blocks are examples of manipulatives that can be used for such exchanges. Games and activities can help provide experience and practice.

Addition and Subtraction Facts

Fact fluency for addition and subtraction is important. I start with addition facts and use that to help students master subtraction facts by pointing out that subtraction is the inverse of addition. This can be done by introducing fact families so that students can see how subtraction facts relate to addition facts. Here are a few examples.

1 + 2 = 3; 3 – 1 = 2	4 + 5 = 9; 9 – 4 = 5
2 + 1 = 3; 3 – 2 = 1	5 + 4 = 9; 9 – 5 = 4

I ask students to work on adding 1 to each single-digit number and then progress to adding 2 and 3. Work could continue to include adding each of the other single digits.

0 + 1	0 + 2	0 + 3
1 + 1	1 + 2	1 + 3
2 + 1	2 + 2	2 + 3
3 + 1	3 + 2	3 + 3
4 + 1	4 + 2	4 + 3
5 + 1	5 + 2	5 + 3
6 + 1	6 + 2	6 + 3
7 + 1	7 + 2	7 + 3
8 + 1	8 + 2	8 + 3
9 + 1	9 + 2	9 + 3

Counting On

When students begin working on addition, I teach them to count on. This ties in with and uses their ability to count. As an example, for 7 + 2 students would count on two numbers beyond 7. So, starting with 7, they would count 7, 8, 9 to find that $7 + 2 = 9$.

Doubles and Doubles Plus One

At some point, students should learn doubles.

$1 + 1 = 2$	$2 + 2 = 4$	$3 + 3 = 6$	$4 + 4 = 8$	$5 + 5 = 10$	$6 + 6 = 12$
$7 + 7 = 14$	$8 + 8 = 16$	$9 + 9 = 18$	$10 + 10 = 20$	$11 + 11 = 22$	$12 + 12 = 24$

Once they learn doubles, students can easily use what they know to do problems where doubles plus 1 will work. Most students will be able to do doubles and doubles plus 1 in their head.

$2 + 3 = 2 + 2 + 1 = 4 + 1 = 5$ $\quad$ $3 + 4 = 3 + 3 + 1 = 6 + 1 = 7$

Adding Ten

Students should use their ability to count by tens to add 10 to any number. This is also an opportunity to add to or reinforce place value knowledge.

$23 + 10 = 33$ $\quad$ $485 + 10 = 495$

At some point, students may encounter a problem such as $598 + 10$. This problem provides an opportunity for students to begin using what they have learned from exchanging 10 tens for 100 to start regrouping.

Once students can readily add 10 to any number, they should be able to use that to add 9 to any number. Nine is $10 - 1$; to add 9 to any number, all students need to do is add 10 to the number and subtract 1.

$23 + 9 = 23 + (10 - 1) = (23 + 10) - 1 = 33 - 1 = 32$

$485 + 9 = 485 + (10 - 1) = (485 + 10) - 1 = 495 - 1 = 494$

Combinations of Ten

I teach students to recognize number pairs that combine to make 10. This helps students develop the ability to quickly add single-digit numbers in their head. It is also a tool they can use to break up numbers to make number pairs that combine to make 10.

1 + 9	6 + 4
2 + 8	7 + 3
3 + 7	8 + 2
4 + 6	9 + 1
5 + 5	

These examples show breaking up a number to make a number pair that will combine to make 10 and help solve the problem.

$7 + 5 = 7 + (3 + 2) = (7 + 3) + 2 = 10 + 2 = 12$ or

$7 + 5 = (2 + 5) + 5 = 2 + (5 + 5) = 2 + 10 = 12$

Horizontal Addition and Subtraction

Many textbooks will ask students to solve problems in a horizontal format. Counting up for addition and counting down for subtraction is a strategy students can use, but like using the number line, this isn't efficient for working with two- and three-digit numbers. Students should be familiar with adding and subtracting using the horizontal format. After elementary school, students frequently see and work with problems written in horizontal format such as $x^2 + 2x - 3$. I provide experience working on problems in horizontal format but push on to work with a vertical format. From my observation, once students are familiar with the vertical format, most will rewrite a horizontal problem in the vertical format. (An example of using the horizontal format is provided below in the "Subtraction with Regrouping" section.)

The addition and subtraction sections that follow focus on using the vertical format. Two numbers are used in the examples. The format and procedures are the same when adding more than two numbers. I always stress with students the importance of aligning the columns by place value.

Addition

Addition is the act of combining two or more objects or numbers together to form a new set that is more than any of the individual objects or numbers being combined. The parts of an addition problem have names students should become familiar with (see Figure 6-1).

tens	ones	
2	3	→ addend
+ 1	4	→ addend
3	7	→ sum

Figure 6-1

Students need to be familiar with the terms "all together," "put together," "how many in all," "total," and "sum." This helps them recognize when addition is the operation to use.

The examples used here include problems involving three-digit numbers. After working with single-digit numbers, I recommend working with two-digit numbers and then three-digit numbers. As students progress through the grades, this can be expanded to adding numbers with more than three digits.

Addition without Regrouping

I would show and solve a few problems in expanded form and begin working with students on doing problems in regular vertical form. Prior experience with place value enables them to have no difficulty lining the columns up appropriately (see Figures 6-2 and 6-3).

$$\begin{array}{rcl} 325 & = & 3 \text{ hundreds} + 2 \text{ tens} + 5 \text{ ones} \\ \underline{+\,243} & = & \underline{2 \text{ hundreds} + 4 \text{ tens} + 3 \text{ ones}} \\ & & 3 \text{ hundreds} + 6 \text{ tens} + 8 \text{ ones} \quad = 568 \end{array}$$

Figure 6-2

$$\begin{array}{rcl} 325 & = & 300 + 20 + 5 \\ \underline{+\,243} & = & \underline{200 + 40 + 3} \\ & & 500 + 60 + 8 = \quad 568 \end{array}$$

Figure 6-3

Step by step, this is how I walk students through solving this problem in regular vertical form (see Figure 6-4).

1. Write the problem down with place value columns lined up. Initially, I may have the place value words written down and work toward just saying the place value without having the words written.
2. Add the digits in the ones column.
3. Add the digits in the tens column.
4. Add the digits in the hundreds column.
5. *Optional:* With great joy, shout, "I did it! I did it! I did it!"

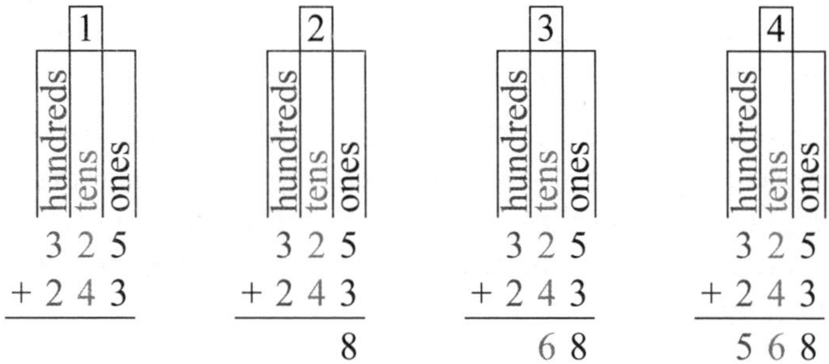

Figure 6-4

This is the same procedure for adding together numbers with any number of digits.

An error students make is not aligning the columns properly. I found it rare for my students to make this error; nevertheless, I always looked to see if they were writing the problems down correctly.

Addition with Regrouping

Students in the second half of second grade should already have plenty of experience exchanging 10 ones for 10, and 10 tens for 100. Along with prior experience and knowledge of place value, students should be able to make the transition from adding without regrouping to adding with regrouping. I make sure to show students the difference between when they need to carry and when they don't need to carry. This is as easy as follows: 1) If the numbers add up to 10 or more, there is a need to carry, and 2) if the numbers add up to less than 10, there is no need to carry.

I show and solve a few problems in expanded form and then begin working with students on doing problems in regular vertical form. I cycle many times through I-We before having students do any independent work. After explaining and showing several examples, I work on and talk through a few problems, having students write down everything I write down as I solve the problems.

Here are some examples of what I might show students before moving on to do problems in regular vertical form (see Figures 6-5, 6-6, and 6-7).

9 7 5	=	9 hundreds	+ 7 tens	+ 5 ones
+ 2 5 8	=	2 hundreds	+ 5 tens	+ 8 ones
	=	11 hundreds	+ 12 tens	+ 13 ones
	=	11 hundreds	+ 12 tens + 1 ten	+ 3 ones
	=	11 hundreds	+ 13 tens	+ 3 ones
	=	11 hundreds + 1 hundred	+ 3 tens	+ 3 ones
	=	12 hundreds	+ 3 tens	+ 3 ones
	= 1 thousand +	2 hundreds	+ 3 tens	+ 3 ones
	= 1 2 3 3			

Figure 6-5

9 7 5	=	9 0 0	+ 7 0	+ 5
+ 2 5 8	=	2 0 0	+ 5 0	+ 8
	=	1 1 0 0	+ 1 2 0	+ 1 3
	=	1 1 0 0	+ 1 2 0 + 1 0	+ 3
		1 1 0 0	+ 1 3 0	+
	=	1 1 0 0 + 1 0 0	+ 3 0	+ 3
	=	1 2 0 0	+ 3 0	+ 3
	= 1 0 0 0 +	2 0 0	+ 3 0	+ 3
	= 1 2 3 3			

Figure 6-6

	thousands	hundreds	tens	ones	
		9	7	5	
	+	2	5	8	
			1	3	ones
		1	2	0	tens
+	1	1	0	0	hundreds
	1	2	3	3	

Figure 6-7

Step by step, I walk students through solving this problem in regular vertical form when regrouping, or carrying is required, as follows:

1. Write the problem down with the place value columns lined up.
2. Add the digits in the ones column. Check to see if carrying is needed. If the sum is less than 10, we don't need to carry. If the sum is 10 or more, we need to carry. In this case, the sum is 13, so we write the 3 in the ones column and the 1 in the tens column above the 7. I would explain that the digit 1 in 13 stands for 1 ten.
3. Add the digits in the tens column, including any number that has been carried over. Check to see if we need to carry. If the sum is less than 10, we don't need to carry. If the sum is 10 or more, we need to carry. In this case, the sum is 13, so we write the 3 in the tens column and the 1 in the hundreds column above the 9. I would explain that in this 13, the 3 stands for 3 tens and the 1 stands for 1 hundred.
4. Add the digits in the hundreds column, including any number that has been carried over. Check to see if we need to carry. If the sum is less than 10, we don't need to carry. If the sum is 10 or more, we need to carry. In this case, the sum is 12, so we write the 2 in the hundreds column and the 1 in the thousands column above where numbers would be if there were any thousands in the original numbers being added. I explain that in this 12, the 2 stands for 2 hundreds and the 1 stands for 1 thousand.

5. Add the digits in the thousands column. In this case, it is only the 1 that was carried over, so it could have just been written in the answer without writing it above in the carry-over row.

Steps 1–4 are shown in Figure 6-8 below.

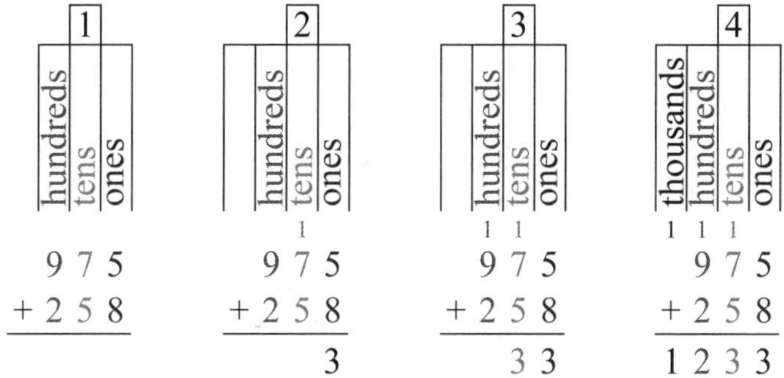

Figure 6-8

In this problem, it is important to note that carrying was required in each step. This is not the case with all problems, so I advise students that it is important to always check to see if carrying is needed. Sometimes students will carry when they don't need to or not carry when they need to. Those are a couple things to watch for with student work. Sometimes, especially early in learning this, some students will write the carry-over number in the answer row instead of up above in the carry-over row. This creates a problem by throwing the place value alignment out of whack.

Occasionally (really rarely), I have seen teachers ask students to solve regrouping problems such as the one in Figure 6-9.

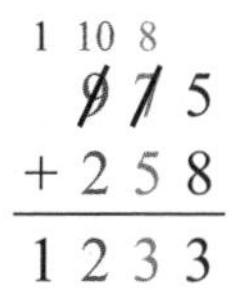

Figure 6-9

Although this works and is mathematically correct, I have never taught students to do it this way. It just doesn't seem efficient to add a 1 to the 7 and write the sum, 8, above the 7, then cross out the 7 and turn around and add 8 and 5. Writing a little 1 above the 7 and then adding 1 + 7 + 5 gets the same result and seems quicker. I save the crossing out for subtraction problems that require borrowing.

Subtraction

Subtraction is the act of finding what is left after a part of a total has been taken away. As with addition, the parts of a subtraction problem have names students should become familiar with (see Figure 6-10).

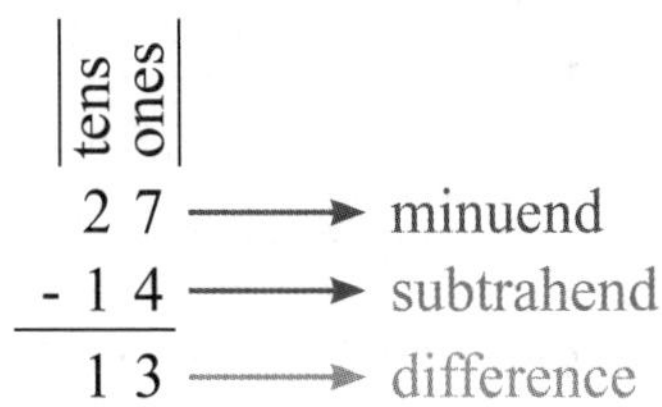

Figure 6-10

Students recognize the need to use subtraction when they are familiar with and see or hear the terms "difference," "less than," "how many more," "minus," "take away," "remain," and "left."

In the early grades and until students begin working with integers, it is important to stress with students that in subtraction of whole numbers the larger number is put on top in vertical form; in horizontal form, the larger number is placed first. In vertical form, starting with the ones column, the bottom digit is subtracted from the digit on top. If the digit on top is smaller than the bottom digit, regrouping or borrowing needs to take place. Trying to take a smaller digit on top from a larger digit on the bottom is a common error students make when learning to subtract with regrouping. I remind students to take the bottom number from the top number rather than the smaller digit from the larger digit.

Subtraction without Regrouping

I show and work on a few problems in expanded form and begin working with the students on problems in regular vertical form (see Figures 6-11 and 12).

I always stress writing the problem down with the larger number on top and the place value columns lined up properly.

$$\begin{array}{rcl} 8\,5\,9 & = & 8 \text{ hundreds } + 5 \text{ tens } + 9 \text{ ones} \\ \underline{-\,6\,0\,9} & = & \underline{6 \text{ hundreds } + 0 \text{ tens } + 9 \text{ ones}} \\ & & 2 \text{ hundreds } + 5 \text{ tens } + 0 \text{ ones } = 2\,5\,0 \end{array}$$

Figure 6-11

$$\begin{array}{rcl} 8\,5\,9 & = & 8\,0\,0 + 5\,0 + 9 \\ \underline{-\,6\,0\,9} & = & \underline{6\,0\,0 + \quad 0 + 9} \\ & & 2\,0\,0 + 5\,0 + 0 = 2\,5\,0 \end{array}$$

Figure 6-12

Step by step, this is how I walk students through solving this problem in regular vertical form.

1. Write the problem down with the larger number on top and the smaller number directly below it with place value columns lined up. Initially, I may have the place value words written down and work toward just saying the place value without having the words written.
2. Subtract the bottom digit from the top digit in the ones column.
3. Subtract the bottom digit from the top digit in the tens column.
4. Subtract the bottom digit from the top digit in the hundreds column.

Steps 1–4 are shown in Figure 6-13 below.

1

	hundreds	tens	ones
	8	5	9
-	6	0	9

2

	hundreds	tens	ones
	8	5	9
-	6	0	9
			0

3

	hundreds	tens	ones
	8	5	9
-	6	0	9
		5	0

4

	hundreds	tens	ones
	8	5	9
-	6	0	9
	2	5	0

Figure 6-13

Subtraction with Regrouping

I stress with students how to properly write a subtraction problem down. Problems are written in the same fashion for problems requiring regrouping as those not requiring regrouping. I make sure to show students the difference between when they need to borrow and when they don't need to borrow. If the digit in the top number is less than the respective digit in the bottom number, borrowing is required. If the digit in the top number is greater than the bottom one, borrowing is not required.

Here is a problem in horizontal format that requires regrouping (see Figure 6-14).

$$
\begin{aligned}
342 - 187 &= (300 + 40 + 2) - (100 + 80 + 7) \\
&= (300 + 30 + 12) - (100 + 80 + 7) \\
&= (200 + 130 + 12) - (100 + 80 + 7) \\
&= (200 - 100) + (130 - 80) + (12 - 7) \\
&= 100 + 50 + 5 \\
&= 155
\end{aligned}
$$

Figure 6-14

Here are two examples (see Figures 6-15 and 16) of the same problem in expanded vertical form. I might show students several problems worked on in this manner before solving problems in regular vertical form or in the form of the standard algorithm.

$$
\begin{array}{rcl}
342 & = & 3 \text{ hundreds} + 4 \text{ tens} + 2 \text{ ones} \\
-\ 187 & = & 1 \text{ hundreds} + 8 \text{ tens} + 7 \text{ ones} \\
\hline
342 & = & 3 \text{ hundreds} + 3 \text{ tens} + 12 \text{ ones} \\
-\ 187 & = & 1 \text{ hundreds} + 8 \text{ tens} + 7 \text{ ones} \\
\hline
342 & = & 2 \text{ hundreds} + 13 \text{ tens} + 12 \text{ ones} \\
-\ 187 & = & 1 \text{ hundreds} + 8 \text{ tens} + 7 \text{ ones} \\
\hline
 & & 1 \text{ hundreds} + 5 \text{ tens} + 5 \text{ ones} = 155
\end{array}
$$

Figure 6-15

$$
\begin{array}{rcccccc}
342 & = & 300+40+2 & = & 300+30+12 & = & 200+130+12 \\
-\ 187 & = & 100+80+7 & = & 100+80+7 & = & 100+80+7 \\
\hline
 & & & & & & 100+50+5 = 155
\end{array}
$$

Figure 6-16

Step by step, this is how I walk students through solving this problem in regular vertical form when regrouping, or borrowing, is required. Starting with the ones column and with each move to the column to the left, I stress the importance of checking to see whether borrowing is going to be required.

1. Subtract the bottom digit from the top digit in the ones column.
2. Subtract the bottom digit from the top digit in the tens column.
3. Subtract the bottom digit from the top digit in the hundreds column.

Here I provide the kind of talking patter I often used when working through problems with my classes. Use your imagination for a good patter on other steps and other example problems.

1. Write the problem down with the larger number on top and the smaller number directly below it with place value columns lined up.

2. In this problem, the 7 looks upstairs and sees that the 2 is less than it. So, the 7 tells the 2 it isn't big enough to take 7 away from it and asks it to get bigger. At this point, the 2 looks to its left and sees 4 tens and asks if it can borrow 1 ten. The 4 is okay with that, saying doing so will make it a 3. That being the case, the 4 gets crossed out and a 3 is written above it. A 1 representing the borrowed 1 ten is written to the left of the 2, making it a 12. "Ah, wonderful!" says the 7. "Now, you are big enough for 7 to be taken away from you"; 7 is taken away from the 12, resulting in a 5, which is written in the answer row in the ones column.
3. Moving to the tens column, check to see if we need to borrow. The 8 is more than the 3, so the 3 will need to borrow from its neighbor on the left, which is a 3. The 3 in the tens column borrows 1 from the 3 in the hundreds column, making the 3 in the hundreds column a 2. The 1 that is borrowed is written next to and to the left of the 3 in the tens column, making it 13. Subtract 8 from 13 and write the resulting 5 in the tens column of the answer row.
4. Since the hundreds column is the last one, we shouldn't need to check to see if we need to borrow. However, if the bottom digit is greater than the top digit in this last column, it means we didn't write the larger number on top. All is okay in this problem. We subtract 1 from the 2 and write the difference of 1 in the hundreds column of the answer row.
5. It's all over but the shouting!

Steps 1–4 are shown in Figure 6-17 below:

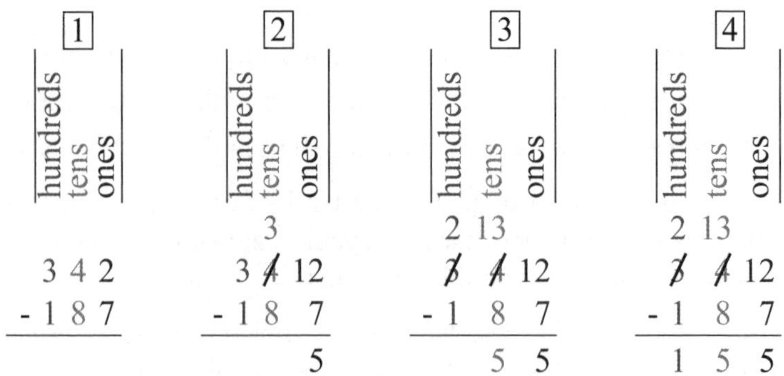

Figure 6-17

In this example problem, borrowing was required in both Steps 2 and 3. That is not the case in all problems, so it is important to always check whether borrowing is needed. If it is not needed, don't do it.

Students may encounter problems that are a bit different from the example used in this section. They may end up working on a problem where they need to borrow for only one column. As students encounter problems with a slight twist, I would sometimes challenge them to use what they know to see if they can complete the problem. If not, I would walk through the problem, calling on students to tell me what to do next. This would involve putting a problem such as the one below on the board, and challenge them to apply what they know. I then see if they can successfully solve it (see Figure 6-18).

$$\begin{array}{r} 3\,2\,4 \\ -\,1\,6\,8 \\ \hline \end{array} \qquad\qquad \begin{array}{rrr} & 11 & \\ 2 & 1 & \\ \not{3} & \not{2} & 14 \\ -\,1 & 6 & 8 \\ \hline 1 & 5 & 6 \end{array}$$

Figure 6-18

Just for Fun. Why don't you do math in the jungle? Because if you add four plus four you get ate!

Why was the subtraction sign elected president? Because he promised to make a difference.

Take 4 and subtract 2. What's left? The opposite of right.

CHAPTER 7: MULTIPLICATION AND DIVISION

SKIP COUNTING

Work on multiplication begins with skip counting in kindergarten. Most students leaving kindergarten at the end of the year can count to 10 by ones and skip count by twos to 20, by fives to 50, and by tens to 100. In Grade 2, students learn to skip count by threes to 30, and in Grade 3, by fours to 40. Skip counting helps students learn and understand multiples as the numbers being counted are multiples of the number being skip counted. The numbers counted are also the multiplication facts for the number being skip counted.

Learning and practicing skip counting in kindergarten, Grade 1, and Grade 2 is a portable oral activity. It can be done as a class while lining up, walking in line, or waiting, or as a filler when opportune times arise.

Topics	Kindergarten	Grade 2	Grade 3	Grade 4	Grade 5
Multiplication	Skip count by twos, fives, tens	Skip count by threes Intro Factor and product Facts for 1, 2, 3, 5, 10 Commutative property Identity property Zero property	Skip count by fours Facts to 10 x 10 Three-digit by one-digit with regrouping Multiplication and division as inverse operations	Mentally by 10, 100, 1,000 Identify multiples and common multiples Multiply two- and three-digit numbers Associative property Estimate	Distributive property Estimate

Dot Cards

In the primary grades, I used dot cards to help students become familiar with multiplication. My dot cards used a strategy of what some refer to as crossed line or intersecting line multiplication. This strategy can be used to multiply multi-digit numbers. I only used a simple form for Grade 1 and 2 students to work on basic multiplication facts. I had 10 sets of large index cards. There was one set for each level of multiplication facts through 10. During the course of a week, every student would complete an activity using the set of cards for that week. Primary teachers are creative and will find a way to effectively use these cards if this is something they feel will work for them in helping their students.

Here's an example of how the cards might help with understanding (see Figure 7-1). For this example, I will use one card presented in a few different ways.

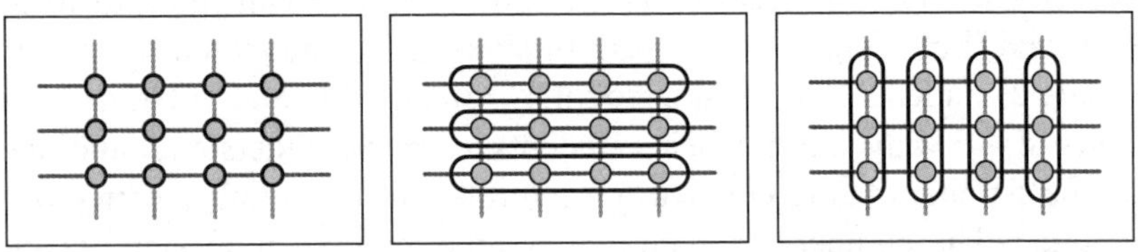

Figure 7-1a, b, and c

The first card is simply 3 red lines intersected by 4 green lines. This can be considered a representation of 3 × 4 or 4 × 3. The answer can be found by counting the number of yellow dots where lines intersect.

The second card shows the 4 yellow dots of intersection along each red line grouped together. There are three groups of four. This represents 3× 4. Rather than counting each individual dot of intersection, counting by fours three times will provide the answer: 4, 8, 12. This can be extended to see the first three multiples of 4. This also represents repeated addition, 4 + 4+ 4 = 12, to reach the answer of a multiplication problem.

The third card shows 3 yellow dots of intersection along each green line grouped together. There are 4 groups of 3. This represents 4 × 3.

Rather than counting each individual dot of intersection, counting by threes 4 times will provide the answer: 3, 6, 9, 12. This can be extended to see the first four multiples of 3. This also represents repeated addition, $3 + 3 + 3 + 3 = 12$, to reach the answer of a multiplication problem. This can also help students see that $3 \times 4 = 4 \times 3$.

MULTIPLICATION WITH ARRAYS

In the early grades, students can be shown multiplication with arrays to help them understand multiplication as repeated addition (see Figure 7-2). Being familiar with multiplication with arrays also helps them better understand *length* × *width* = *area* of a rectangle.

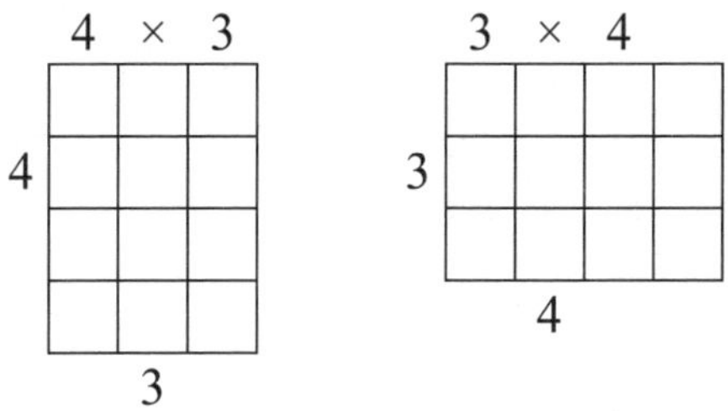

Figure 7-2

Multiplication with arrays is a good way to show students perfect squares (see Figure 7-3).

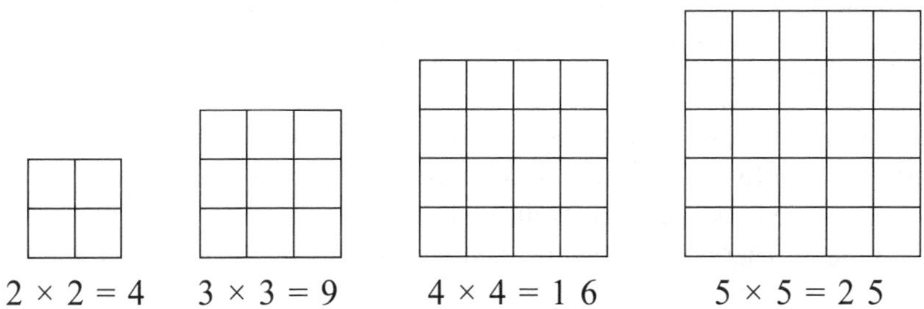

Figure 7-3

MULTIPLICATION FACTS

Students entering Grade 3 able to skip count by twos, threes, fives, and tens will learn to skip count by fours. That means, possibly without realizing it, from skip counting they already know their multiplication facts for 1, 2, 3, 4, 5, and 10. That leaves the facts for 6, 7, 8, and 9 to be learned in Grade 3. Recognizing that 6, 8, and 9 are multiples of 3, 4, and 3, respectively, can make learning the facts for those numbers easier. The facts for 7 are often the hardest for students to learn. From skip counting and knowing the multiplication facts for the other numbers, student will already be familiar with the multiples of 7, except for 7 × 7, which students may not have yet learned as a perfect square.

X	**1**	**2**	**3**	**4**	**5**	**6**	**7**	**8**	**9**	**10**
1	1	2	3	4	5	6	7	8	9	10
2	2	4	6	8	10	12	14	16	18	20
3	3	6	9	12	15	18	21	24	27	30
4	4	8	12	16	20	24	28	32	36	40
5	5	10	15	20	25	30	35	40	45	50
6	6	12	18	24	30	36	42	48	54	60
7	7	14	21	28	35	42	49	56	63	70
8	8	16	24	32	40	48	56	64	72	80
9	9	18	27	36	45	54	63	72	81	90
10	10	20	30	40	50	60	70	80	90	100

A multiplication table can be helpful for skip counting and learning multiples and multiplication facts. It is interesting to see what patterns students will find in it.

Since about three-fourths of my students entered Grade 6 only being proficient with multiplication facts through the threes, I was always happy if my students learned them through the tens. It was icing on the cake if students knew their facts through the twelves by the end of Grade 6.

Fact Fluency

Students need to be proficient with their math facts. Once basic addition facts are learned, the facts for the inverse operation of subtraction should be easy. The same is true about multiplication facts and the facts for the inverse operation of division. Quick and accurate recall of basic math facts is math fact fluency. Math fact fluency is key to "computation fluency, or the ability to quickly and accurately compute the answer to a given number sentence" (Davies, 2022).

The Rocket Math (2019) website provides three reasons why math fact fluency is important: 1) students with math fact fluency make fewer errors, 2) math fact fluency makes learning math easier, and 3) students who have developed math fact fluency enjoy math and always complete their work.

Most students need to work to develop math fact fluency. The work can be fun and rewarding. Most math work is written work, so it is important for students to develop fluency with writing numbers. Without writing fluency, students may be slowed down in their work. To help with this, I give students timed drills on skip counting.

In addition to timed drills on skip counting, I give sixth-graders timed drills to develop fluency with their basic multiplication facts.

Timed Drills

Timed drills can be done for addition and subtraction facts in first and second grade. When I taught those grades, I used a drill program that students enjoyed, and it had the elements of what I look for in a drill program. These elements include having students set goals, consistent daily practice of intense but short duration, leveled practice focusing on one fact or skill at a time, self-checking, immediate feedback, record keeping to see progress, and the ability to move up to a new level when an established proficiency goal has been reached.

For sixth grade, I ask students to do two drills. One is a written fluency timed drill with skip counting. The other is a timed drill with basic multiplication facts. Students work on the same level for skip counting and multiplication facts. In other words, if a student is working on their multiplication facts for 4, they will do a skip counting drill and multiplication drill for the fours.

The two kinds of drills can be done in the classroom with the whole class or a parent can give these drills at home. If they are well planned with the elements mentioned earlier, both drills can be done from start to finish in 10–15 minutes.

Timed Drill: Written Fluency with Skip Counting. Students write their skip counting numbers in this drill as fast as they can to see how many numbers they can write in a minute. I encourage students to do the skip counting with the voice in their head as they write the numbers down. The drill is repeated three times. It is typical for students do get their best score on the second round.

I ask students to keep a record sheet and papers for this drill in a folder. When the class is ready, I tell them to start. I time them for one minute as they write their skip counting numbers as many times as they can. If a student is working on their fours, their work may look something like this but with more lines completed.

4 8 12 16 20 24 28 32 36 40

4 8 12 16 20 24 28 32 36 40

4 8 12 16 20 24

At the end of the minute, when told to stop, students count up how many numbers they have written. They record this on their record sheet. This cycle is completed three times. I encourage students to set a goal for the day before the drill starts. This is not a handwriting activity, so I don't expect the writing to be neat, but it should be legible.

Timed Drill: Multiplication Facts. At the beginning of the year in sixth grade, I assess students on their multiplication facts. The assessment is used to place them on an initial level of multiplication facts. A few students each year might test out and be able to do other things during drill time.

Each student has a folder. The folder contains a record sheet, a fresh fact worksheet, old used worksheets, and an answer key for their current worksheet. When the class is ready, I tell them to start. Everyone works on their problems until I say stop at the end of the time. Five minutes is usually the allotted amount of drill time. Depending on the class and grade level, I may alter the amount of time. At the end of the time, students get their answer key out of their folder and check their work. If they complete the drill paper with 100% correct, they see me and get set up for the next level. Those students who do not move up a level get a fresh worksheet and put it in their folder to be ready for the next day.

I had multiplication fact worksheets for each level (i.e., a worksheet for 2× facts, 3× facts, 4× facts, through 10×). When a student completed problems through the tens, their final worksheet had mixed facts. Each worksheet had 100 problems.

Work on fact fluency is important. This set-up worked for me and my students. I encourage other teachers to find a way that works for them to help students develop fact fluency.

In my sixth-grade classes, multiplication drills would continue until mid-December. By that time, about three-fourths of the students had exited the drills.

From what I have observed, if students don't have their multiplication facts mastered before entering junior high or middle school, they may never master them.

Multiplication Terminology. It is important for students to know the terminology for multiplication. The number being multiplied is called the multiplicand, and the number doing the multiplying is called the multiplier. The result, or answer, of a multiplication problem is called the product (see Figure 7-4).

24	multiplicand	24	factor
× 7	multiplier	× 7	factor
168	product	168	product

Figure 7-4

Just for Fun. 24 (multiplicand) × 7 (multiplier) = 7 (multiplicand) × 24 (multiplier). Is it possible the commutative property creates a multiplication identity crisis?

The multiplicand and multiplier are both factors of the product and are often referred to as factors. Does that resolve any possible identity crisis?

The sign of multiplication is an × and is read as "multiplied by" or "times." The problem 2 × 5 might be read as "two multiplied by five" or as "five times two" (or "two times five"). Although I might say five times two, I wouldn't say "times two by five." That's just me. If I'm timing something, I use a stopwatch.

Multiplying a Multi-digit Number by a One-Digit Number. Multiplying a multi-digit number by a one-digit number or by another multi-digit number is easier when students understand place value, are able to apply properties, and know basic multiplication facts. It's ideal if students are able to multiply multi-digit numbers by one-digit numbers by the end of Grade 3 and multiply two multi-digit numbers by the end of Grade 4.

I generally start by showing how to multiply multi-digit numbers by one-digit numbers in horizontal form using the distributive property. I then show the same problem in a more standard form. The example problem in Figure7-5 requires carrying when done in the standard form. Prior to showing and having students work on problems that require carrying, I ask students to work through ones that don't require carrying.

Below is an example of multiplication using the distributive property.

7×24

$7 \times (20 + 4) = (7 \times 20) + (7 \times 4) = 140 + 28 = 168$

The standard method of multiplication is shown in Figure 7-5.

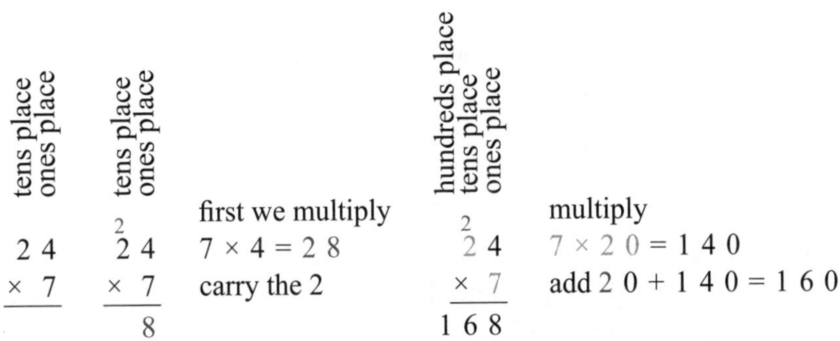

tens place
ones place
24
× 7
tens place
ones place
2
24
× 7
8
first we multiply
7 × 4 = 2 8
carry the 2
hundreds place
tens place
ones place
2
24
× 7
1 6 8
multiply
7 × 2 0 = 1 4 0
add 2 0 + 1 4 0 = 1 6 0

Figure 7-5

The digit in the ones place of the multiplier is multiplied by the digit in the ones place of the multiplicand and then by the digit in the tens place of the multiplicand. The products are combined (added) to get the final product, as shown.

Multiplying a Multi-digit Number by Another Multi-digit Number. Multiplying a multi-digit number by another multi-digit number is similar to multiplying a multi-digit number by a one-digit number except that more steps are involved. First, the multiplicand is multiplied by the ones digit of the multiplier. This can be called the first partial product. Second, the multiplicand is multiplied by the tens digit of the multiplier. This can be called the second partial product. Third, the multiplicand is multiplied by the hundreds digit of the multiplier. This can be called the third partial product. If there are more digits, this process continues until the multiplicand has been multiplied by every digit in the multiplier. The partial products should be written with the digit on the right end in the same place value column as the digit used as a multiplier. The partial products are added to obtain the final product. Students understand the procedure by seeing worked examples and using them as a guide in solving the problems themselves.

Figure 7-6 below shows two ways to carry out the procedure. The second method uses zeros inserted as place holders to help students correctly line up the digits in the partial products.

hundreds thousands	ten thousands	thousands	hundreds	tens	ones	place value
			1	1		carrying
			4	3	3	multiplicand
		×	2	3	4	multiplier
		1	7	3	2	first partial product
	1	2	9	9		second partial product
8	6	6				third partial product
1	0	1	3	2	2	product

			1	1	
			4	3	3
		×	2	3	4
		1	7	3	2
	1	2	9	9	0
8	6	6	0	0	
1	0	1	3	2	2

Figure 7-6

The above method is the form and procedure of the standard algorithm for multi-digit multiplication. During scratch-paper work, I work through problems with the students and, as much as possible, explain in detail what I am doing and why. I emphasize how knowledge of place value helps to carry out the multiplication efficiently using the standard algorithm. If I asked my students if they would like to work the above problem using repeated addition, I know what their answer would be. And it would be loud, passionate, and quick.

I do show and explain other forms for multiplying. In doing so, I make every effort to convey their relation to the how and why of the standard algorithm. Figure 7-7 illustrates another way I might show students the same problem.

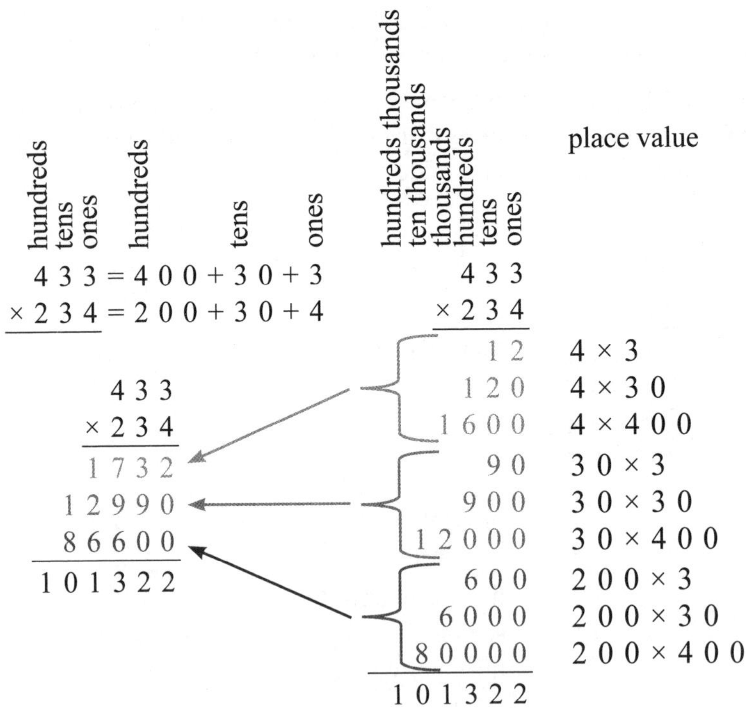

Figure 7-7

Division

Only a few students each year would enter my sixth-grade classroom moderately proficient with long division. For that reason, I would provide some instruction and practice during our scratch-paper time. I would basically start with a review of what division is and then work from one-digit divisors up to multi-digit divisors. Work would progress fairly quickly as students would begin to understand how and why the standard algorithm works. At some point, I would include problems with remainders and decimals.

Most students became very proficient with long division. Many students had learned the partial quotient or big seven method for doing division. Some of these students eventually transitioned to using the standard algorithm when they understood it, had some practice with it, and appreciated its efficiency. A few students had difficulty with the transition and usually dropped back to using the method they were first taught and had practiced for two years. The difference was noticeable. Given a division problem with a three-digit divisor, students using the standard algorithm would successfully complete the problem in one to two minutes, whereas students using an alternative method would take up to 10 minutes or never complete the problem at all.

Understanding that division is repeated subtraction is great, but it is not an efficient way to solve problems, especially when working with larger numbers.

$12 \div 3 = ?$

$12 - 3 = 9$

$9 - 3 = 6$

$6 - 3 = 3$

$3 - 3 = 0$

3 was subtracted from 12 four times. So, $12 \div 3 = 4$.

My students never liked the idea of trying to solve problems such as $174410 \div 214$ using repeated subtraction. And I never subjected them to such problems.

When teaching long division using the standard algorithm, I would show expanded version examples with explanations so students could see and understand how the standard algorithm works and takes advantage of place value knowledge. I stayed away from using or teaching the partial quotients method for three reasons: 1) I didn't find it to be efficient, 2) I could never figure how it worked well with decimals, and 3) I am not sure how well it would work for students later when they would need to do long division with polynomials.

The standard algorithm for division works well when dividing polynomials. Although elementary students won't face these kinds of problems, solid grounding in long division should prepare them well for when they later encounter these kinds of problems, one of which is shown in Figure 7-8:

$$\frac{x^6 + 2x^4 + 6x - 9}{x^3 + 3}$$

$$
\begin{array}{r l}
 & x^3 + 2x - 3 \\
\cline{2-2}
x^3 + 3 & \big) \; x^6 + 2x^4 + 0x^3 + 6x - 9 \\
 & -\underline{(x^6 \qquad + \; 3x^3)} \\
 & \qquad 2x^4 - 3x^3 + 6x - 9 \\
 & \qquad -\underline{(2x^4 \qquad + 6x)} \\
 & \qquad\qquad - 3x^3 \qquad - 9 \\
 & \qquad\qquad -\underline{(- 3x^3 \qquad - 9)} \\
 & \qquad\qquad\qquad 0 \qquad\quad 0
\end{array}
$$

Figure 7-8

I place great importance on students learning to be successful with long division. In a study by Siegler et. al. (2012), student knowledge and success with division at the elementary level was an early predictor of high school math achievement. An online article (Prodigy, 2020) mentions, but doesn't cite, a French study saying long division is "a synthesis of all arithmetic knowledge."

It is helpful for students to already have the skills and knowledge employed in long division. They should know their basic facts as well as know and be able to work with place value, rounding, estimation, division, multiplication, and subtraction. Remainders call for an understanding and ability to work with fractions and decimals. Students can use their knowledge of division as the inverse of multiplication in two big ways. First, they already know basic division facts when they relate multiplication facts to division. Example: $3 \times 4 = 12$ so $12 \div 3 = 4$ and $12 \div 4 = 3$. Second, they can multiply the divisor by the quotient to check their work. Example: $12 \div 3 = 4$ so $3\times 4 = 12$. It also helps if students are familiar with the rules of divisibility.

Students should be familiar with four terms related to division. The "divisor" is the number doing the dividing. As shown in Figure 7-9, the "dividend" is the number being divided or that the divisor goes into. The "quotient" is the answer to a division problem. A "remainder" is what's left over if the divisor does not go into the dividend evenly.

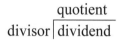

$$\begin{array}{r} \text{quotient} \\ \text{divisor}\overline{\smash{)}\text{dividend}} \end{array}$$

Figure 7-9

Rules of Divisibility. Rather than throwing all the rules of divisibility out to students at once, I like to have students become familiar with each rule as they learn the respective multiplication facts for the given number.

Divisibility means you are able to divide a number evenly. If a number is divisible by 2, it is an even number. If it is not divisible by 2, it is an odd number. I sometimes would tell students that just because a number looks strange does not mean it is odd.

A number is divisible by	If
2	The last digit is even. 5**8** Yes 4**7** No
3	The sum of its digits is divisible by 3. 744 7+4+4 = 15 15÷3 = 5 Yes 476 4+7+6=17 17÷3 = $5\frac{2}{3}$ No
4	The last two digits are divisible by 4. 7,582,348 48÷4 = 12 Yes 724,597 97÷4 = $24\frac{1}{4}$ No

A number is divisible by	If
5	The last digit is 0 or 5. 356**5** Yes 251**3** No
6	The number is divisible by 2 and 3. 120 divisible by 2 (even number) 1+2+0 = 3 3÷3 = 1 Yes 302 divisible by 2 3+0+2 = 5 5÷3 = $1\frac{2}{3}$ No
7	Double the last digit. Subtract it from the other digits. If the result is divisible by 7, the original number is also. 679 double 9 is 18 67–18 = 49 49÷7 = 7 Yes 902 double 2 is 4 90–4 = 86 86÷7 = $12\frac{2}{7}$ No
8	The last three digits are divisible by 8. 765,816 816÷8 = 102 Yes 88,888,888,083 083÷8 = $10\frac{3}{8}$ No
9	The sum of its digits is divisible by 9. 6912 6+9+1+2 = 18 18÷9 = 2 Yes 3152 3+1+5+2 = 11 11÷9 = $1\frac{2}{9}$ No
10	The last digit is 0. 24**0** Yes 20**4** No
An odd number is not divisible by an even number.	

The Long Division Build Up. With most things, especially math, I found my students did best when I would break things down to the easiest and simplest level. Once I would do that, I then started building up by having students do more difficult problems and adding more complexity. Below I show a progression of example problems from the easy and simple to the more difficult and complex. These are the kinds of problems I would cycle through I-We-You during scratch-paper time. I would show an example of a problem in expanded form and the same problem worked through with the kind of talking points I use when solving a problem for students.

Simply put, the steps in long division are divide, multiply, subtract, bring down, and repeat or record the remainder. As I talk through problems with students, it becomes evident there are often other things to consider. Division can be represented in fraction form, with the ÷ sign, or in what I call a division box. The examples here will use a division box.

Progression of Example Problems

Simple One-Digit Divisor Division. If students practice using what they know about the rules of divisibility, they will recognize the dividends in Figure 7-10 as divisible by the divisor:

$$2\overline{)4} \qquad 2\overline{)4\ 6} \qquad 5\overline{)8\ 3\ 4\ 5}$$

Figure 7-10

A good time to introduce remainders would be after going through the above problems. Students should then be able to determine if the dividends in the problems in Figure 7-11 are divisible by the divisor:

$$2\overline{)5} \qquad 2\overline{)4\ 3} \qquad 5\overline{)8\ 3\ 4\ 2}$$

Figure 7-11

If they aren't divisible, there will be a remainder. Remainders are written as a remainder, a fraction, or a decimal (see Figure 7-12):

$$\begin{array}{r} 2\,\text{r}1 \\ 2\overline{)5} \\ -4 \\ \hline 1 \end{array} \qquad \begin{array}{r} 2\tfrac{1}{2} \\ 2\overline{)5} \\ -4 \\ \hline 1 \end{array} \qquad \begin{array}{r} 2\ .5 \\ 2\overline{)5\ .0} \\ -4 \\ \hline 1\ \ 0 \end{array}$$

Figure 7-12

Dividing two or more digit dividends by a two-digit divisor is shown in Figure 7-13:

$$1\ 5\overline{)8\ 0} \qquad 2\ 4\overline{)1\ 6\ 8} \qquad 1\ 6\overline{)5\ 5\ 8\ 4}$$

Figure 7-13

Dividing by Three- and Four-Digit Divisors. Even though I may show examples and ask students to work through a few problems with four-digit divisors, I would call it quits at three-digit divisors. My sixth-grade students liked the challenge of seeing if they could complete four-digit divisor problems in under a minute or between one and two minutes. Sample problems are shown in Figure 7-14:

125 ⟌ 38250   744 ⟌ 9672   2514 ⟌ 37710

Figure 7-14

Talking through Long Division. The same problem is presented here in expanded form and in standard algorithm form. I work through the problem both ways for students. In working through problems, I talk through the steps, including my thought process. As I talk through the steps with the students, I constantly ask questions of the class. Such questions are along the lines of "What should I do next?" "Why would I do that?" or "Why did I do that?" "Can anyone explain why doing that works?" I will "talk" through working the standard algorithm example here without asking the questions.

The questions I ask while working through a problem follow Figure 7-15 below.

	5		
	10		
800 + 10 + 5 = 815	800	815	
	214 ⟌ 174410	214 ⟌ 174410	
800 × 214 = 171200	-171200	-1712↓	8 × 214 = 1712
	3210	321	
10 × 214 = 2140	-2140	-214↓	1 × 214 = 214
	1070	1070	
5 × 214 = 1070	-1070	-1070	5 × 214 = 1070
	0	0	

Figure 7-15

174410÷214: First, does 214 go into 1? No.

Does it go into 17? No.

Does it go into 174? No, because 174 is a smaller number.

So, it must go into 1,744.

The first digit of my answer will go directly above the 4 that is in the hundreds place of the dividend.

(One alternative to the steps above is to first consider that 214 is a three-digit number and start off by asking if it goes into the first three digits of the dividend, in this case 174. It doesn't, so I would look at 1,744. Two hundred fourteen rounds to 200, and 1,744 rounds to 1,700, so how many times does 200 go into 1,700 or how many times does 2 go into 17? Eight times, so I write 8 above the 4 in the hundreds place of the dividend.)

After determining that 214 must go into 1,744, I then multiply 214 by 8 and get 1,712, which I write directly below 1,744 and subtract it. If I had tried 9, the result of multiplying 214 by 9 would be larger than 1,744, letting me know 9 is too large. If I had tried 7, the result of multiplying 214 by 7 would give 1,498. Subtracting 1,498 from 1,744 gives me 246. Two hundred forty-six is larger than 214, and 214 will go into it once, letting me know that 7 won't work, but 1 more than 7, or 8, will work.

I subtract 1,712 from 1,744 and get 32. I make sure to write the digits in the correct place value column. I've now taken care of 1,744 and need to bring the 1 down that is in the tens column. The 1 is brought down to make the 32 become 321.

How many times does 214 go into 321? I know it will go into it at least once. What about twice? 214 × 2 is 428, which is more than 321, so I know it goes in only once.

I write the 1 in the tens column above the 1 in the dividend. Then, I multiply 214 by 1 and get 214, which is written directly under 321, and subtracting it gives 107.

Two hundred fourteen won't go into 107, so I bring down the 0 to have the 107 become 1,070. How many times will 214 go into 1,070? Two hundred fourteen rounds to 200, and 1,070 rounds to 1,000. Two hundred goes into 1,000 five times or 2 goes into 10 five times. The 5 is written in the ones column directly above the 0 in the dividend.

Now, I multiply 214 by 5 and get 1,070, which is written directly below 1,070 and subtracted. This subtraction gives a 0, letting us know there is no remainder.

The quotient, or answer, is 815. I can multiply 214 by 815 to check my answer.

Short Division. People talk about long division but I never hear them talk about short division. There are problems that are referred to as short division. Short division takes some short cuts and doesn't do the work below that is done in long division. Typically, short division is used when there is a single-digit divisor. Short division lends itself to students being able to write their answer down as they do the work in their heads. Short division problems can be written with a regular division box or with the division box turned upside done as in the example below.

$$\begin{array}{r} 1\,7\,5 \\ 5\overline{)8\,7\,5} \\ \underline{5} \\ 3\,7 \\ \underline{3\,5} \\ 2\,5 \\ \underline{2\,5} \\ 0 \end{array} \qquad\qquad \begin{array}{r} {}^{3}\;{}^{2} \\ 5\,\underline{|\,8\,7\,5} \\ 1\,7\,5 \end{array}$$

Five goes into 8 once. Write 1 down to start the answer. Subtract 5 from 8 and write the 3. Then, bring down the 7 so that we have 37. Five goes into 37 seven times. Write the 7 down next to the 1 in the answer. Multiply 5 and 7 and subtract the resulting 35 from 37 to get 2, which is written to make the 5 become 25. Five goes into 25 five times. Five is written next to the 7 in the answer to show the answer is 175. This is work many of my students would be able to do in their heads.

So, there's long division and short division. What about medium division? If there is short division, does that mean tall division is out there somewhere? Since there is big seven division, why isn't there small two division or medium five division?

Just for Fun. It's all fun and games until someone divides by zero.

The minus sign tried to explain to the plus sign how multiplication works, … but he only understood sum of it.

What tool is best suited for math? Multi-pliers.

CHAPTER 8. BITS, PIECES, AND WHOLE THINGS WITH EXTRA PARTS: FRACTIONS, DECIMALS, AND PERCENTS

Fractions

Students in the primary grades should have lots of concrete and manipulative experience with fractions and advance to visuals like boxes, pie diagrams, and number lines. By the end of second grade, students should be comfortable and familiar with $\frac{1}{2}$, $\frac{1}{3}$, $\frac{1}{4}$, $\frac{1}{5}$, $\frac{1}{6}$, $\frac{1}{8}$ and $\frac{1}{10}$. In fifth and sixth grade, they should master the four operations with fractions, mixed numbers, and improper fractions.

Parts of a Fraction. A fraction is a number that represents a part of a number or group. Like other things, it is important for students to learn the terminology used with fractions. The bottom number of a fraction is called the "denominator" (see Figure 8-1). It tells us how many equal pieces a whole thing has been divided into. The top number is called the "numerator." It tells us how many of those equal pieces are being talked about. A line separates the numerator and denominator. I have always referred to that line as the fraction bar. I've recently learned the fraction bar actually has a name: the "vinculum." Had I known that while I was teaching, I doubt I would have used it or taught it. After my sixth-grade students knew the terms "numerator" and "denominator," they occasionally would catch me saying "carburetor" and "terminator" or "upstairs" and "downstairs."

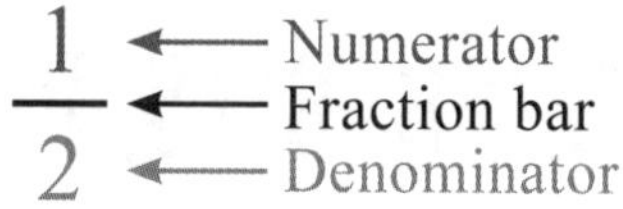

Figure 8-1

Types of Fractions. Students need to become familiar with the following terms describing types of fractions. Any fraction with a numerator of 1 is a unit fraction. If the numerator is smaller than the denominator, it is a proper fraction: $\frac{3}{8}$. An improper fraction has a numerator greater than the denominator: $\frac{7}{4}$. A mixed number has a whole number and a proper fraction: $2\frac{1}{2}$.

Diagrams can show students the progression of increasing the number of equal parts until the fraction becomes an improper fraction (see Figure 8-2).

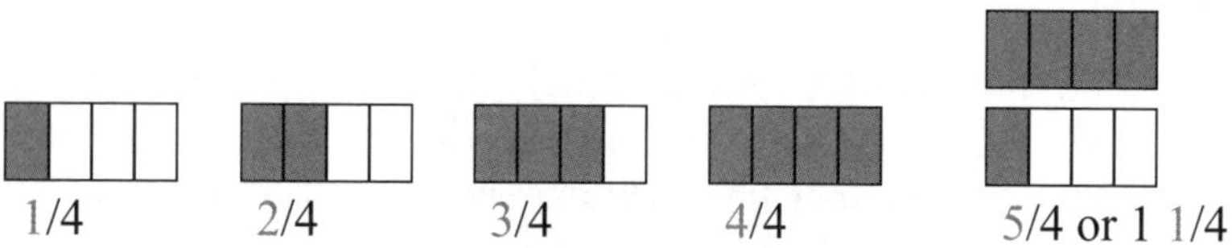

Figure 8-2

Converting Fractions. Students need to be able to convert improper fractions to mixed numbers and mixed numbers to improper fractions. At the elementary level, these conversions are taught in isolation. A lot of teachers will stress that you always have to convert an improper fraction to a mixed number. I don't stress that. I teach students to convert mixed numbers to improper fractions to multiply and divide. Later, when students work with complex algebra equations, they will find it easier to work with an improper fraction than a mixed number.

To convert an improper fraction to a mixed number, you first divide the numerator by the denominator. The whole number answer gets written down, and the remainder is written as the numerator above the denominator.

$$\frac{11}{4} = 2\,\frac{\square}{4}; \quad \frac{11}{4} = 2\,\frac{3}{4}$$

Here's how I would walk and talk the process, writing down the steps shown above that I do in my head:

To convert the fraction $\frac{11}{4}$, the numerator (11) is greater than the denominator (4), so it is an improper fraction. To convert, I divide the numerator, 11, by the denominator, 4, and get 2 with a remainder of 3. The 2 is a whole number and is written down as such. The denominator is still 4. The remainder, 3, is what is left of the 11 in 11/4 and is the numerator in the fraction part of the mixed number $2\,\frac{3}{4}$. It's usually easier to convert if the improper fraction has been reduced to simplest terms.

In a way, the process is reversed to convert a mixed number to an improper fraction. To convert a mixed number to an improper fraction, I teach students to multiply the whole number by the denominator and add the product to the numerator. The answer gets written as the numerator over the original denominator. Figure 8-3 shows an example.

$$2\frac{3}{4} = \frac{2x4+3}{4} = \frac{11}{4} \qquad 2\frac{3}{4} = \frac{11}{4}$$

$$2 \times 4 = 8 \quad 8 + 3 = 11$$

Figure 8-3

To convert the $2\frac{3}{4}$ to an improper fraction, you multiply the whole number, 2, by the denominator, 4. The result is 8, which represents 8/4, which is equivalent to the whole number 2. Since there is already $\frac{3}{4}$ in the mixed number, the $\frac{8}{4}$ gets added. This is easily done by adding the 8 to the 3 and writing 11 as the new numerator over the denominator 4 to get 11/4.

Reducing Fractions: The Diet Process. Reducing a fraction to its simplest terms usually makes it easier to work with the fraction. When a fraction is reduced to simplest terms, the value remains the same. There are a number of methods to reduce fractions. Three of the more common methods are common factors, greatest common factor (GCF), and prime factors. I focus on the first two methods since I teach reducing fractions well before teaching prime factorization.

The GCF method is more efficient than the common factor method. Using the GCF method, I ask students to look for the GCF for the numerator and denominator. So, to reduce $\frac{24}{36}$, students need to first find the GCF of 24 and 36.

- The factors of 24 are 1, 2, 3, 4, 6, 8, 12.
- The factors of 36 are 1, 2, 3, 4, 6, 9, 12, 18.
- The common factors of 24 and 36 are 1, 2, 3, 4, 6, and 12. The GCF is 12.

$$\frac{24}{36} = \frac{24 \div 12}{36 \div 12} = \frac{2}{3}$$

The numerator and denominator are both divided by the GCF, 12. This results in $\frac{2}{3}$, which can't be reduced any further.

The GCF method works fine when students readily find the GCF. I show students that if they don't readily find the GCF to look for any common factor and start with that. In this case, I readily see both the numerator and denominator are even numbers, so I could start by dividing by two. I also see that both are divisible by three and decide to start with that.

$$\frac{24}{\mathbf{36}} = \frac{24 \div 3}{\mathbf{36} \div 3} = \frac{8}{\mathbf{12}} = \frac{8 \div 4}{\mathbf{12} \div 4} = \frac{2}{\mathbf{3}}$$

Dividing both the numerator and denominator by the common factor of 3 gives us $\frac{8}{12}$. Here's where students need to be careful. Some will think they are done. I always let students know to check if the fraction can be further reduced. In this case, 8 and 12 are both divisible by 4. The same result as before, $\frac{2}{3}$, is obtained. I show students they could solve this by dividing both top and bottom by 2 and continuing on. The key is to divide by a common factor and continue until the fraction is in its lowest or simplest terms.

Equivalent Fractions. Fractions having the same value are equivalent even if they don't have the same numerator and denominator. I teach students to make equivalent fractions by multiplying or dividing the top and bottom by the same number. Since adding or subtracting the same amount from the top and bottom will not work, I try to catch and correct that when students make such an attempt.

$$\frac{1}{2}=\frac{2}{4};\quad \frac{1}{2}=\frac{8}{16}$$

Here, two equivalent fractions for $\frac{1}{2}$ were found. The first was found by multiplying the top and bottom by 2. The second was found by multiplying the top and bottom by 8; $\frac{1}{2}$, $\frac{2}{4}$ and $\frac{8}{16}$ are equivalent fractions.

I teach students two ways to determine if two fractions are equivalent. The first way is to reduce both fractions to simplest terms. If that results in the same fraction, the two fractions are equivalent. Let's take $\frac{2}{3}$ and $\frac{488}{732}$ and see if they are equivalent. $\frac{2}{3}$ is already in simplest terms. $\frac{488}{732}$ reduces down to $\frac{2}{3}$. They are equivalent. If when using this method, the two fractions don't reduce to the same fraction, they are not equivalent.

The second way is to cross multiply. If the products of cross multiplying are the same, the two fractions are equivalent. If the products are different, the fractions are not equivalent. Let's look at how cross multiplying works. In this case, the results are the same, so the two fractions are equivalent.

$$\frac{2}{3}\ ?\ \frac{488}{732};\quad \frac{3\times 488=1464}{2\times 732=1464};\quad 1464 = 1464\ so\ \frac{2}{3}=\frac{488}{732}$$

(Cross multiplication is also discussed in Part II Chapter 13, as well as why it works.)

Whole Numbers as Fractions. There are times when students need to represent a whole number as a fraction. I show students two types of examples of whole numbers written as fractions. The first is when they need to write a fraction that represents 1. I remind students the bottom number tells us how many equal pieces something is divided into, and the top number is how many of those pieces are being talked about. If the top number and the bottom number are the same, one whole item is being talked about. The number 1 can be written as a fraction an infinite number of ways as long as the top and bottom number are the same. Here are examples of fractions representing 1: $\frac{1}{1}$, $\frac{2}{2}$, $\frac{3}{3}$, $\frac{4}{4}$, $\frac{5}{5}$, $\frac{36}{36}$, $\frac{824}{824}$, etc. Any whole number can be written as a fraction by writing it as the numerator over a denominator of 1. Examples: $5 = \frac{5}{1}, 12 = \frac{12}{1}$.

Common Denominators. Comparing, ordering, adding, and subtracting fractions are all easier to do if the fractions have common denominators. In other words, all having the same denominator. Finding the least common denominator and multiplying the top and bottom of each fraction by the denominator of the other are the two methods I teach students to use for finding common denominators.

Finding the least common denominator is the same as finding the least common multiple of the two or more denominators being used. Here's how I show students how to find the least common denominator for $\frac{1}{5}$, $\frac{1}{6}$, and $\frac{1}{15}$. First, list multiples of each denominator.

Multiples of 5 are 5, 10, 15, 20, 25, **30**, 35, …

Multiples of 6 are 6, 12, 18, 24, **30**, 36, 42, …

Multiples of 15 are 15, **30**, 45, 60, …

Second, identify the smallest number that is a multiple for all three denominators. In this case, the least common denominator of $\frac{1}{5}$, $\frac{1}{6}$, and $\frac{1}{15}$ is 30. Next, write an equivalent fraction for each with the common denominator. Identify the number to multiply the denominator by to get 30 and then multiply that by the numerator to find the numerator of an equivalent fraction with a denominator of 30.

$\frac{1}{5} = ?\frac{\square}{30}$ $5 \times 6 = 30$ or divide 30 by 5 and get 6 so $\frac{1}{5} = \frac{6}{30}$.
$\frac{1}{6} = ?\frac{\square}{30}$ $6 \times 5 = 30$ or divide 30 by 6 and get 5 so $\frac{1}{6} = \frac{5}{30}$.
$\frac{1}{15} = ?\frac{\square}{30}$ $15 \times 2 = 30$ or divide 30 by 15 and get 2 so $\frac{1}{15} = \frac{2}{30}$.

Sometimes the lowest common denominator is the common denominator found by multiplying the denominators of the two fractions. For example, the fractions $\frac{2}{5}$ and $\frac{1}{2}$ have a common denominator of 10, as shown.

$$\frac{2}{5} = \frac{2 \times 2}{5 \times 2} = \frac{4}{10}; \quad \frac{1}{2} = \frac{1 \times 5}{2 \times 5} = \frac{5}{10}$$

Using this method to find a common denominator of $\frac{2}{5}$ and $\frac{1}{2}$, the numerator and denominator of $\frac{2}{5}$ are multiplied by the denominator, 2, of the other fraction. The numerator and denominator of $\frac{1}{2}$ are multiplied by 5, the denominator of $\frac{2}{5}$. In doing this, an equivalent fraction for $\frac{2}{5}$ is found that has the same denominator as an equivalent fraction for $\frac{1}{2}$ without any change in value.

Other times, the denominator of one fraction is a multiple of the other; for example, $\frac{1}{3}$ and $\frac{5}{6}$. In this case, only one fraction needs to be converted: $\frac{1}{3}$ is converted to $\frac{2}{6}$. These situations are also addressed in the chapter on adding and subtracting algebraic fractions in Part III, Chapter 15.

It is also important to note that students don't always need or have to use the least common denominator. Sometimes it is quicker or easier to use a common denominator by multiplying the top and bottom of each fraction by the denominator of the other, even if a lower common denominator can be found. For example, the least common denominator for $\frac{3}{4}$ and $\frac{5}{6}$ is 12. A common denominator is 24. If 24 is the common denominator, the fractions would then be $\frac{18}{24}$ and $\frac{20}{24}$.

Comparing, Ordering, and Rounding Fractions. Students well versed and comfortable working with equivalent fractions and common denominators seem to find comparing and ordering fractions fairly easy. The first step in comparing fractions is making sure the fractions have a common denominator. If the fractions all have the same denominator, comparing the numerators is all that needs to be done. The fractions can then be arranged in order according to the size of the numerators.

I do teach students to check and see if they can use a quick and easy way to compare two fractions with different denominators. The quick and easy way is to see if the fractions are greater or less than $\frac{1}{2}$. If they are both greater or both the same, then common denominators need to be used. If they are not the same, the one that is more than $\frac{1}{2}$ is greater than the one less than $\frac{1}{2}$. Students only need to divide the denominator by 2 and see if the numerator is more or less than that. They usually are able to do this in their head.

Students can easily round a mixed number to the nearest whole number using what they learn about comparing fractions. If the fraction is $\frac{1}{2}$ or greater, it is rounded up to the next whole number. If the fraction is less than $\frac{1}{2}$, the fraction is dropped, and the existing whole number stays the same.

Fractions can be compared and rounded by converting to decimals. The methods described above are usually quicker than converting to decimals and require less work. My students always seemed to appreciate being able to do things in ways requiring the least amount of work.

Adding and Subtracting Fractions. I try to break this down and streamline it as much as possible for students. I teach students to add and subtract fractions with like denominators first. Once they are able to do that, then adding and subtracting fractions with unlike denominators becomes easy to do when they use equivalent fractions having a common denominator.

Adding and subtracting mixed numbers can be a bit trickier for students. I show students two ways. The easier way for some is to convert the mixed numbers to improper fractions having a common denominator and then add or subtract. If this results in an improper fraction, students need to convert to a mixed number. Another way involves dealing with the fraction part of the mixed number and then the whole number part. With addition and subtraction, the fraction parts need to have a common denominator. If the fraction part of the minuend, or the larger top number, is less than the fraction part of the subtrahend, or smaller bottom number, then borrowing is needed. An example of this is shown in the subtraction section.

Adding Fractions. I teach students to add fractions using the following three steps.

Step 1. Make sure the fractions have common denominators. If they don't, then find and use equivalent fractions with common denominators.

Step 2. Add the numerators of the fractions with common denominators.

Step 3. Simplify to lowest terms if necessary.

Here is an example showing $\frac{2}{5}$ and $\frac{1}{2}$ being added. These fractions don't have the same denominator, so I show students how to find and use their equivalent fractions with common denominators, $\frac{4}{10}$ and $\frac{5}{10}$.

$$\frac{4}{10}+\frac{5}{10}=\frac{4+5}{10}=\frac{9}{10}$$

This example is written horizontally. I find it easier for students to work with these kinds of problems horizontally because they can readily see whether the denominators are common. Initially, I show students how I rewrite the problem adding the two numerators, 4 + 5, over one denominator, as shown above.

At first, students tend to add the two denominators together, 10 + 10, so rewriting the problem this way helps prevent that. The result of this addition is $\frac{9}{10}$. Students should see if the result is in lowest terms. If it is, as in this case, they don't need to do anything. If it isn't, they will need to simplify.

Subtracting Fractions. I teach subtracting fractions after teaching addition. The steps are the same. Once the fractions have common denominators, students need to compare the fractions to make sure they are subtracting a fraction of lesser value from one of greater value.

Step 1. Make sure the fractions have common denominators. If they don't, then find and use equivalent fractions with common denominators.

Step 2. Subtract the numerators of the fractions with common denominators.

Step 3. Simplify to lowest terms if necessary.

This example shows $\frac{2}{5}$ being subtracted from $\frac{1}{2}$. Equivalent fractions with common denominators have already been determined. $\frac{2}{5}=\frac{4}{10}\cdot\frac{1}{2}=\frac{5}{10}$

The fractions $\frac{4}{10}$ and $\frac{5}{10}$ are compared to make sure $\frac{1}{2}$ is greater than $\frac{2}{5}$. It is, so $\frac{2}{5}$ can be subtracted from $\frac{1}{2}$ by subtracting $\frac{4}{10}$ from $\frac{5}{10}$.

$$\frac{5}{10}-\frac{4}{10}=\frac{5-4}{10}=\frac{1}{10}$$

Adding and Subtracting Mixed Numbers. As mentioned earlier, I teach students two ways to add and subtract mixed numbers. The first is to convert the mixed numbers to improper fractions with common denominators. I work these problems horizontally when showing students how to do it. The second way requires showing students problems worked vertically by lining up the whole number parts and the fraction parts, as is done with place value columns when adding or subtracting multi-digit whole numbers. One example of vertical addition requiring carrying and one example of vertical subtraction requiring borrowing is shown here. While talking students through the process, I encourage them to recall and apply their knowledge of carrying and borrowing from regular addition and subtraction.

Adding mixed numbers with carrying is shown in Figure 8-4:

13 3/4		= 13 3/4	Write fractions with common denominators.
+ 5 1/2	= 5 +(1/2 × 2/2)	= 5 2/4	
		= 18 5/4	Add whole numbers. Add fractions.
		= 18 +1 1/4	Convert 5/4 to mixed number.
		= 19 1/4	Carry the whole number 1. Fraction part is in lowest terms.

Figure 8-4

Subtracting mixed numbers with borrowing is shown in Figure 8-5:

54 1/4	= 53 + 1 1/4 = 53 + 5/4	= 53 5/4	1/4 borrows 1 from 54. Rewrite 54 1/4 as 53 5/4.
− 24 1/2	= 24 +(1/2 × 2/2)	= 24 2/4	Rewrite with common denominator of 4.
		= 29 3/4	Subtract fractions. Subtract whole numbers.

Figure 8-5

Multiplication of Fractions. Multiplying fractions always seemed to come easy for my students since they could solve the problems without having to find and use common denominators. When students would need to multiply mixed numbers, I found it easiest to ask them to convert the mixed numbers to improper fractions, multiply, and then convert the result back to a mixed number and simplify the fraction part as needed.

Students would use their ability to multiply fractions to divide fractions. They would only need to tend to one thing. That one thing is addressed in the section on dividing fractions.

I taught my students three easy steps for multiplying fractions.

Step 1. Multiply the numerators.

Step 2. Multiply the denominators.

Step 3. Simplify to lowest terms if necessary.

Students are used to getting larger numbers when multiplying whole numbers and may be surprised when that doesn't happen with fractions. Explanations and visuals of what happens when they multiply fractions is helpful. I work toward students understanding they are finding a fraction of a fraction. In the first example in Figure 8-6 below, $\frac{1}{4}$ of $\frac{1}{3}$ or $\frac{1}{3}$ of $\frac{1}{4}$ is found to be $\frac{1}{12}$. In the second example, $\frac{1}{4}$ of $\frac{2}{3}$ or $\frac{2}{3}$ of $\frac{1}{4}$ is $\frac{1}{12}$, which reduces to $\frac{1}{6}$. Two visual examples are shown in Figure 8-6, followed by the same problems worked in the way I would show students:

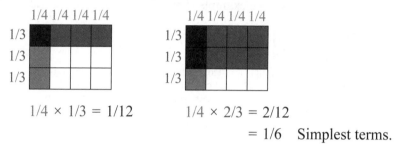

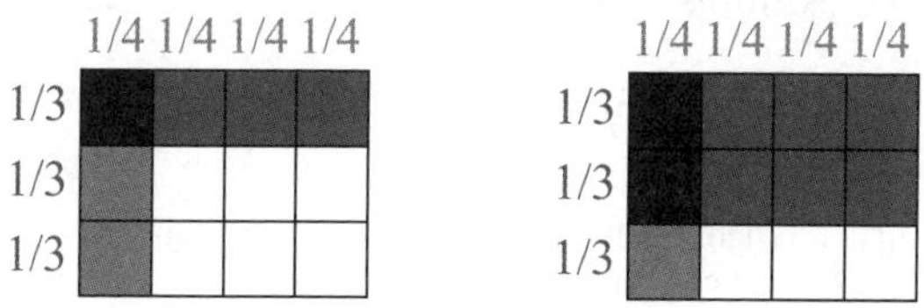

1/4 × 1/3 = 1/12

1/4 × 2/3 = 2/12

= 1/6 Simplest terms.

Figure 8-6

And here is the problem worked out: $\frac{1}{4} \times \frac{1}{3} = \frac{1\times1}{4\times3} = \frac{1}{12}$ $\frac{1}{4} \times \frac{2}{3} = \frac{1\times2}{4\times3} = \frac{2}{12} = \frac{1}{6}$

I like to show students how to simplify across fractions before multiplying. Students appreciate this method, especially when dealing with large numerators and denominators. Here's how I show students what is going on with this simplification:

$$\frac{1}{4} \times \frac{2}{3} = \frac{1\times2}{4\times3} = \frac{1\times2}{3\times4} = \frac{1}{3} \times \frac{1}{2} = \frac{1}{6}$$

The commutative property allows us to reverse the factors in the denominator. That is, 4 × 3 = 3 × 4. I used that principle to show the progression in Figure 8-7 below. This progression shows students that they can go right to Step c to complete the problem while skipping Steps a and b. The top and bottom both get divided by a common denominator. In this example, they are divided by 2:

$$\overset{a}{\frac{1 \times \cancel{2}\,1}{3 \times \cancel{4}\,2}} = \overset{b}{\frac{1 \times \cancel{2}\,1}{2\,\cancel{4} \times 3}} = \overset{c}{\frac{1}{2\,\cancel{4}} \cdot \frac{\cancel{2}\,1}{3}} = \overset{d}{\frac{1}{2} \cdot \frac{1}{3}} = \frac{1}{6}$$

Figure 8-7

Division of Fractions. I have found that showing visual models of division with fractions often confuses students. For that reason, I try to show a simple model or two so they can see what is going on before moving on. Here's a simple example showing $\frac{3}{4} \div \frac{1}{8} = 6$:

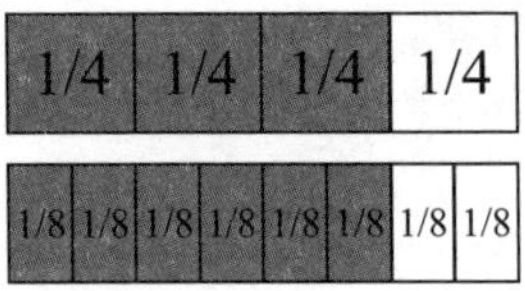

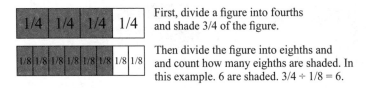

First, divide a figure into fourths and shade 3/4 of the figure.

Then divide the figure into eighths and and count how many eighths are shaded. In this example. 6 are shaded. 3/4 ÷ 1/8 = 6.

Figure 8-8

Figure 8-8 shows that there are 6 one-eighths in $\frac{3}{4}$.

These are the steps I teach for dividing fractions.

Step 1. Write the original division problem.

Step 2. Rewrite the problem as a multiplication problem using the reciprocal of the divisor.

Step 3. Multiply the numerators.

Step 4. Multiply the denominators.

Step 5. Simplify to lowest terms if necessary.

First, students need to know what a reciprocal is. If a fraction is multiplied by its reciprocal, the product is 1. The reciprocal of a fraction is simply written by reversing the numerator and denominator; that is, $\frac{4}{3}$ is the reciprocal of $\frac{3}{4}$. $\frac{3}{4} \times \frac{4}{3} = \frac{12}{12} = 1$. Next, it is important for students to see why multiplying by the reciprocal of the divisor works. I found that most of my students were satisfied with an explained example but usually didn't remember it even though they could successfully divide fractions.

Using variables and then numbers, I show students examples of how flipping the divisor and multiplying works to divide a fraction by a fraction. As I show it, I provide an explanation of what is being done and why it can be done. Unfortunately, most textbooks I have looked at and used never provided much, if any explanation. My hunch is that most students are never shown much more than to flip the divisor and multiply. My students got the whole kit and caboodle from me. There are a number of ways to show how to derive the flip and multiply. This is the way I showed students. It was easier for me to show and explain than some other ways.

$$\frac{a}{b} \div \frac{c}{d} = \frac{\frac{a}{b}}{\frac{c}{d}} = \frac{\frac{a \cdot d}{b \cdot c}}{\frac{c \cdot d}{d \cdot c}} = \frac{\frac{a \cdot d}{b \cdot c}}{1} = \frac{a \cdot d}{b \cdot c}$$

Here is an example using numbers.

$$\frac{3}{4} \div \frac{1}{8} = \frac{\frac{3}{4}}{\frac{1}{8}} = \frac{\frac{3 \cdot 8}{4 \cdot 1}}{\frac{1 \cdot 8}{8 \cdot 1}} = \frac{\frac{3 \cdot 8}{4 \cdot 1}}{1} = \frac{3 \cdot 8}{4 \cdot 1} = \frac{24}{4} = 6$$

Upon seeing how this works, my students were always happy to skip the extra work and jump right into working problems like this.

$$\frac{3}{4} \div \frac{1}{8} = \frac{3}{4} \cdot \frac{8}{1} = \frac{24}{4} = 6$$

(Another demonstration is provided in Part II Chapter 8, for seventh grade.)

DANCING DECIMALS

In my sixth-grade class, students learned when and how to do the decimal dance. Decimals do a one-step, two-step, or promenade dance. Before going to the decimal ball, students need to understand and be able to work with decimals.

Most students already have experience working with money written with a decimal. I build on this experience to help students understand that the decimal part represents a fraction of a whole.

Numbers with decimals may have a whole number to the left of the decimal point and the fractional part to the right of the decimal point. I build on student knowledge of place value and expand it to decimals. Figure 8-9 shows decimal place value going out five places. Figure 8-10 shows the same number written out in expanded form and how the number should be read.

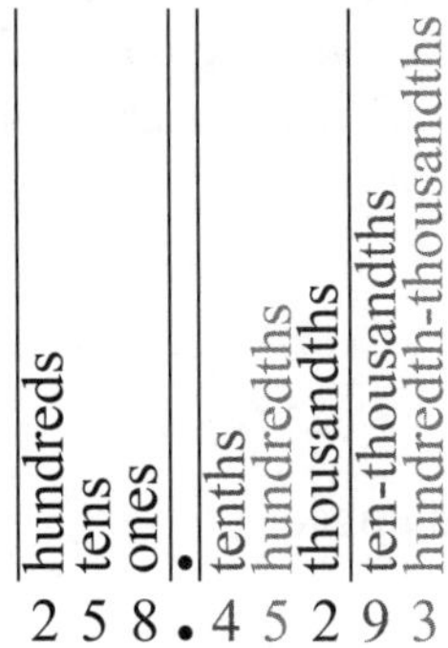

Figure 8-9

200.00000	two hundred
50.00000	fifty
8.00000	eight
.40000	four tenths
.05000	five hundredths
.00200	two thousandths
.00090	nine ten-thousandths
+ .00003	three hundred-thousandths
258.45293	

Read as two hundred fifty eight and forty-five thousand, two hundred ninety three hundred-thousandths

Figure 8-10

Like a lot of people, my students had a tendency to read a decimal number, such as 3.45, in one of the following ways: 1) three dot four five; 2) three point four five; or 3) three decimal point four five.

I explain the correct way to read a decimal number. In addition to always reading decimals correctly, I insist students do so as well. Doing so shows an understanding of place value. In my classroom, 3.45 is read as "three and forty-five hundredths." The decimal point is read as "and." It is not read as "dot," "point," or "decimal point." I am picky about this. I let my students know, and they met my expectation.

It can be helpful to model decimals for students. Modeling can also help students see and understand the connection between fractions and decimals. Student understanding of fractions should help them see and understand with each place to the right of the decimal, the value gets smaller and smaller. Here, the fraction is shown above its decimal equivalent.

$$\frac{1}{10}, \frac{1}{100}, \frac{1}{1000}, \frac{1}{10{,}000}, \frac{1}{100{,}000}$$

$$0.1,\ 0.01, 0.001,\ 0.0001,\ 0.00001$$

Comparing and Ordering Decimal Numbers. I teach students to write decimal numbers with the decimal points lined up to help them compare. If there are whole number parts, it is easy to compare unless they all have the same whole number. If the decimal numbers all have the same whole number or have no whole number, then students look first to see which has the highest number in the tenths place. As needed, they would continue on to the next place to the right. An example is shown in Figure 8-11, followed by the steps:

Compare	Order, largest to smallest
0.25	0.3
0.208	0.2999
0.2999	0.25
0.3	0.208

Figure 8-11

1. Since there is no whole number present, look first at the tenths place. Three is the largest, so 0.3 is the largest of these four numbers.
2. The rest all have 2 in the tenths place, so look at the hundredths place. Nine is the largest, and 0.2999 is larger than the remaining two numbers.
3. Next, compare the 5 and 0 in the hundredths place of the remaining two numbers. Five is the largest, so 0.25 is larger than 0.208. Yes, 8 is greater than 5 but the place value it is in (i.e., the thousandths place) makes it lesser than the value 5 represents.

Adding and Subtracting Decimal Numbers

I teach my students that adding and subtracting decimal numbers is done just like regular addition and subtraction once the numbers are written properly. The numbers should be written with the decimal points directly lined up with the decimal point in the number above. If the numbers didn't have the same number of decimal places, my students found it helpful to add zeros to the right as needed so the numbers all had the same number of decimal places. Adding zeros to the right of a decimal number does not change the value. The decimal point in the answer is placed in line with the other decimal points. Here's an example of addition and subtraction with decimal numbers (see Figure 8-12):

$$\begin{array}{r} {\scriptstyle 1\ \ \ \ \ 1\ 1\ \ \ \ \ } \\ 32.59831 \\ 4.01900 \\ +\,47.20010 \\ \hline 83.81741 \end{array} \qquad\qquad \begin{array}{r} {\scriptstyle 0\ \ 12\ 11\ \ } \\ \not{1}.\not{3}\not{1}4 \\ -\,0.580 \\ \hline 0.734 \end{array}$$

Figure 8-12

Multiplying Decimal Numbers

I teach students to multiply decimal numbers as if they were multiplying whole numbers. After adding and subtracting decimals, students have a tendency to set up a multiplication problem by lining the decimals up. Instead of doing that, I teach them to line the numbers up on the right side as if they didn't have decimals. The numbers get multiplied without regard for the decimals. Once students have a product, they place the decimal, so there is the same number of decimal places as the total decimal places in the factors. For easy practice, I show multiplication problems and ask students to determine how many decimal places the answer needs to have.

In the example in Figure 8-13, the first factor has three decimal places, and the second factor has one decimal place for a total of four decimal places. That determines there will be four decimal places in the product. I tell students to let the problem tell them how many places should be in the answer.

```
  4.2 2 2 ←— 3 decimal places ⎫  3 + 1 = 4
×   2.3 ←— 1 decimal place   ⎭  4 decimal places in the product
 1 2 6 6 6
 8 4 4 4
 9.7 1 0 6 ←— 4 decimal places
```

Figure 8-13

Dividing Decimal Numbers

Decimals start dancing when I show students how to divide decimal numbers. To start, I show students how to divide a decimal number by a whole number. No dancing needed for that. Then, I move on to whole numbers or decimal numbers by a decimal number divisor. I show students that you start by rewriting the divisor by multiplying it by a multiple of 10 to make it a whole number. The dividend needs to be multiplied by the same multiple of 10. I remind students that they previously learned about division being written in fraction form and getting an equivalent fraction by multiplying the top and bottom by the same factor. I show them this is what they are doing when they multiply the divisor and dividend by the same multiple of 10.

In the example in Figure 8-14, the divisor and dividend are multiplied by 100 to make the divisor be the whole number 6. In doing so, the decimal dances two steps to the right in both numbers. Since both the divisor and dividend end up being whole numbers, there is no need to write a decimal point.

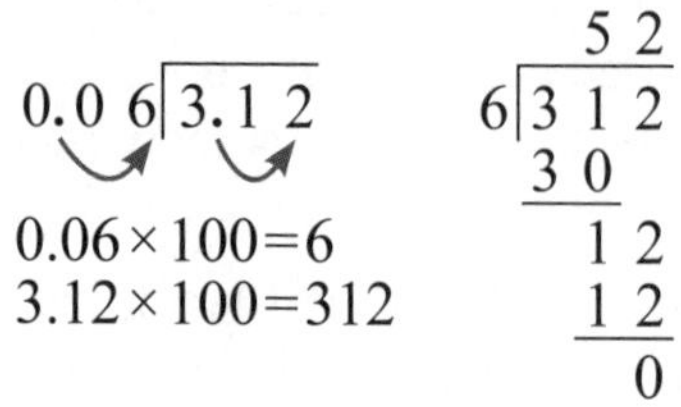

Figure 8-14

Figure 8-15 is an example of the remainder written as the decimal portion of the quotient:

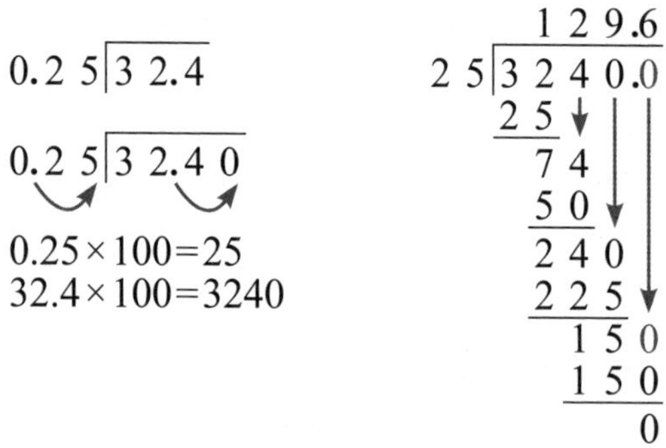

Figure 8-15

In this problem, the decimals each dance two steps to the right. In order to do that, a zero has to be added to the end of 32.4, making it 32.40 with two decimal places so the dance step can be completed. Twenty-five does not go into 3,240 evenly, so a decimal followed by a zero is added. This does not change the value of 3,240. Some problems may need additional zeros added. The decimal point gets placed in the quotient directly above the decimal point in the dividend.

Decimal Dance Steps. The decimal dances whenever a number is multiplied or divided by a multiple of 10. The decimal dances as many steps as there are trailing zeros in the multiple of 10. The decimal dances to the right in multiplication and to the left in division.

41 × 100 = 4100. The decimal dances two steps to the right.

41 ÷ 100 = 0.41. The decimal dances two steps to the left.

Students have been doing the decimal dance without realizing it when they have added or removed trailing zeros when multiplying or dividing by a power of 10. Any whole number, such as 41, can be written with a decimal point and any number of zeros following the point. Forty-one can be written as 41.00, and the value of 41 remains the same. Typically, whole numbers are not written with a decimal point, but if it is helpful, we can use a decimal point.

41 × 100 can be rewritten as 41.00 × 100

41.00 × 100=4100. The decimal dances two places to the right.

When the decimal dances two places to the right, the 4 that was previously in the tens place is now in the thousands place, and the 1 that was in the ones place is now in the hundreds place. The number 4,100 is now 100 × 41. So, when we multiply by 100 the decimal shifts, or dances, two places to the right to obtain the product.

The decimal dances to the left in division.

41 ÷ 100 can be rewritten as 41.0 ÷ 100

41.0 ÷ 100 = 0.41. The decimal dances two steps to the left.

Since 41 does not have a visible decimal point, we can add one by writing 41.0. When the decimal dances two places to the left, the 4 that was previously in the tens place is now in the tenths place, and the 1 that was in the ones place is now in the hundredths place. When we divide 41 by 100, the decimal shifts, or dances, two places to the left to obtain the quotient of 0.41.

Metamorphic Math: Fraction-Decimal-Percent Conversions

I teach my students to use their super powers to change back and forth among fractions, decimals, and percents without changing values. At this point, students already have the prerequisite skills and knowledge for working with fractions and decimals. I teach them to use the skills and knowledge they already have to learn and use conversion basics.

Percents. Percents may need to be taught before jumping into conversions. Percent means parts per 100. Students relate to this because they know there are 100 cents in a dollar, 100 cents = $1.00. I teach students to think of a percent as a fraction that always has a denominator of 100, like $\frac{15}{100}$, and is written 15% with a % symbol instead of the denominator.

Fraction to Decimal. A fraction is converted to a decimal by dividing the numerator by the denominator (see Figure 8-16).

$$\frac{3}{8} = .375$$

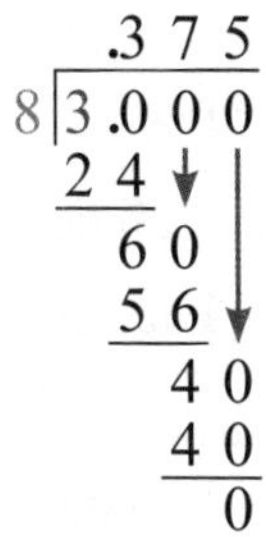

Figure 8-16

Decimal to Fraction. To convert the decimal to a fraction, the decimal is first written as a fraction over 1. The decimal point in the numerator needs to take as many dance steps to the right as needed to make it a whole number. To keep up with it, the decimal point in the denominator needs to take the same number of steps to the right. To convert .375 to a fraction, it is first written as a fraction with .375 over 1. The decimal point in both the numerator and denominator needs to dance three steps to the right. To get the decimal point to take those dance steps, the numerator and denominator are multiplied by 1,000. In the example in Figure 8-17, the result can be reduced to $\frac{3}{8}$.

$$\frac{.375}{1} = \frac{.375}{1,000} = \frac{375}{1000} = \frac{375 \div 125}{1000 \div 125} = \frac{3}{8}$$

Figure 8-17

Decimal to Percent. To convert a decimal to a percent, the decimal point does the two-step, always to the right. Once the two steps to the right are taken, a % symbol is added at the end (see Figure 8-18).

$$.375 = 37.5\%$$

Figure 8-18

Percent to Decimal. Converting a percent to a decimal requires that the decimal point in the percent dance two steps to the left. Once the two steps are taken, the % symbol is dropped (see Figure 8-19).

$$37.5\% = .375$$

Figure 8-19

Fraction to Percent. To convert a faction to a percent, the fraction is first converted to a decimal, and the decimal is then converted to a percent.

Percent to Fraction. To convert a percent to a fraction, the percent is first converted to a decimal, and the decimal is then converted to a fraction.

Practical Percents. Students need to be able to figure the percent of some number. I teach my students two ways to do this. The first way is to convert the percent to a decimal number and multiply the number by the decimal number. To find 35% of 120, 35% is converted to a decimal number and then multiplied by 120.

$$35\% \text{ converts to } 0.35 \rightarrow 0.35 \times 120 = 42$$

In this way, we find that 35% of 120 is 42.

The second way is to use what they know about equivalent fractions—that they are proportional—to find the percent of some number. The percent is first converted to a fraction. That fraction is then made part of an equivalent fraction equation. For example, to find 35% of 120, 35% is first converted to the fraction $\frac{35}{100}$, which becomes part of the equation below:

$$\frac{35}{100} = \frac{\square}{120} \text{ or } \frac{35}{100} = \frac{x}{120}$$

For the fraction $\frac{35}{100}$, 35 is the part, and 100 is the whole. In the second fraction, we are finding the part that is called x. Because it is the part, it is placed in the numerator; the whole is 120, representing 100% and is placed in the denominator.

Since the fractions are equivalent, they can be cross multiplied:

$\frac{35}{100} = \frac{x}{120}$ cross multiplied becomes $35 \times 120 = 100x$; then $4200 = 100x$.

Both sides of the equation are then divided by 100.

$\frac{4200}{100} = \frac{100x}{100}$. This results in $42 = x$ or $x = 42$.

The equivalent fraction method results in the same answer as the first method (i.e., 35% of 120 is 42).

I give students simple practice problems to find 10% of different numbers. What is 10% of 23? What is 10% of 435? What is 10% of 8,672? Initially, students solve these problems on paper. After some practice, I ask students to solve these kinds of problems in their head. This sets them up for being able to figure tips.

Tips: Useful Information. I always made it a point to teach my sixth-grade students how to figure tips of 10%, 15%, and 20% in their heads. First, they would learn to mentally determine 10% of an amount by moving the decimal one place to the left. Let's say they have gone out to eat and the bill comes to $32.25. Moving the decimal one place to the left gives $3.225, which is 10% of $32.25. To make it easier to do mentally, I ask students to round to the nearest tenth. So, for a bill of $32.25, a 10% tip can be figured as $3.20 and only be a few cents off. Once 10% is figured, finding 15% and 20% is easy. Twenty percent is double 10%, so double $3.20 is $6.40, which gives us the amount for a 20% tip. For a 15% tip, we figure that 15% is 10% plus 5%. Five percent is half of 10%, so we take half of $3.20, which is $1.60, and add it to $3.20. In this way, we find that $4.80 would be the amount for a 15% tip.

After I taught students this technique, some would later happily share with me that they figured the tip for their parents when they ate at a restaurant. I was told by some that their parents didn't know how to figure a tip and would just leave extra money on the table.

The above is only a "percent" of what I do with students in sixth grade as we work with percents. Barry covers percents well in the section on

Grade 7. He goes into greater depth in working with and solving problems with percents. (See Part II Chapters 16–20.)

Ratios and Proportions

Students need to become familiar with ratios and proportions before moving on to seventh and eighth grade, where these topics become an important part of their work. In Part II, Barry thoroughly addresses both ratios and proportions and takes students from using tape diagrams to solving ratio problems using algebra. (See Part II, Chapters 11–13.) Ratios and proportions are briefly addressed here as taught in sixth grade.

Ratios. In sixth grade, students learn that a ratio compares two numbers by division. If there are 32 blue flags and 20 red flags, I ask students to compare the number of blue flags to red flags, which is 32 to 20. This is comparing a part of the whole (i.e., total amount) to another part of the whole. This is a part-to-part ratio.

I then ask them to compare the number of blue flags to the total number of flags. For this, students need to add 32 and 20 to get the total, 52; they can now say there are 32 blue flags to a total of 52 flags. This is a part-to-whole ratio; that is, comparing a part of the whole to the whole, or total. I teach students ratios can be written three ways, 32 to 20, 32:20, and $\frac{32}{20}$. Except when asked to present a ratio differently, I encourage students to write ratios in fraction form. I show students that ratios can be worked with like fractions. The ratio $\frac{32}{20}$ is equivalent to $\frac{16}{10}$ and $\frac{8}{5}$.

Rates. I teach students that rates are ratios comparing quantities measured in different units. An example would be a car going 300 miles on 12 gallons of gas. This would be written as 300 mi/12 gal. To find how far the car would go on 1 gallon of gas, I ask students to find an equivalent ratio for 1 gallon. A quantity compared to a unit of one is a unit rate. The unit rate for 300 mi/12 gal is 25 mi/1 gal.

Proportions. A proportion is an equation stating two ratios are equal. Stated differently, if two ratios are equivalent, they are a proportion. The ratio $\frac{32}{20}$ is equivalent to $\frac{16}{10}$ so 32/20 =16/10 is a proportion. Like fractions, ratios can be cross multiplied to see if they are equivalent.

$\frac{32}{20} ? \frac{16}{10}$; cross multiply $\frac{20 \times 16 = 320}{32 \times 10 = 320}$; $320 = 320$ so $\frac{32}{20} = \frac{16}{10}$ and is a proportion.

Cross products are the products of cross multiplying. This process can be used to find a missing factor in a proportion. If we know $\frac{32}{20} = \frac{\text{something}}{10}$, and we need to find the "something," we can use cross products:

$$\frac{32}{20} = \frac{x}{10} \rightarrow 20x = 32x20 \rightarrow 20x = 320 \rightarrow \frac{20x}{20} = \frac{320}{20} \rightarrow x = 16$$

In using this method to find a missing factor in a proportion, I take the opportunity to teach students how to solve one-step problems like this by dividing both sides of the equations by the same number, as shown above. I tell students this is an algebra skill they will learn in seventh and eighth grade. That gets their attention.

OTHER STUFF

About midway through the second half of sixth grade, there's more stuff I like my students to work with that I feel they need before moving on to seventh grade. Students need to know about integers and how to use the four operations with positive and negative integers. They learn about factors and how to find factor pairs for a composite number as well as how to use a factor tree for prime factorization. Students learn how to solve for the variable in one- and two-step problems as well as working with exponents. Like percents, ratios, and proportions, Barry addresses some of these additional topics well in Part II (seventh grade) and in Part III (eighth-grade algebra), so these topics aren't presented here.

Just for Fun. I like to order my fractions with ranch dressing on the side.

Why should you never argue with decimals? Decimals always have a point.

There's a fine line between a numerator and a denominator… but only a fraction would understand.

Why was the fraction worried about marrying the decimal? Because she would have to convert.

References for Part I

Davies, R. (2022). Why math fact fluency matters. *Differentiated Teaching*. www.differentiatedteaching.com/math-fact-fluency-matters/

Milgram, R. J. (2005). The mathematics pre-service teachers need to know (p. 476). Unpublished manuscript prepared for U.S. Department of Education. www.scribd.com/document/118887879/James-Milgram-Math-Preservice-teachers-need-to-know

Prodigy. (2020, Oct. 16). How to do long division: A simple step-by-step guide with pictures. *Prodigy*. www.prodigygame.com/main-en/blog/how-to-do-long-division/

Rocket Math. (2019, July 9). The ultimate guide to math fact fluency. www.rocketmath.com/2019/07/09/guide-developing-math-fact-fluency/

Siegler, R. S., Duncan, G. J., Davis-Kean, P. E., Duckworth, K., Claessens, A., Engel, M., Susperreguy, M. I., & Chen, M. (2012). Early predictors of high school mathematics achievement. *Psychological Science*, 23(7), 691–697. (Original DOI: 10.1177/0956797612440101)

PART II.A
SEVENTH-GRADE MATH

The examples of lessons in this part (II.A and B) draw from a sequence of topics that I and other teachers with whom I've worked believe make the most mathematical and logical sense.

I've noticed many math books for seventh grade start with ratios and proportions. I believe this order is followed because it is also the order of topics in the Common Core Math Standards for that grade. There is nothing in the standards that prescribes that order, and I feel certain that teachers will not be punished for using a sequence different from what appears in either the Common Core standards or in the textbooks aligned with them.

I introduce algebraic expressions and equations prior to ratios and proportions, one of the largest topics in seventh-grade math. I do this because I believe students need the appropriate background and tools with which to better understand and operate with the concepts to solve problems.

What follows is the scope and sequence for a regular seventh-grade math course (Math 7) and an accelerated one. Part II.A is primarily regular Math 7, though there are some topics included that are also in the accelerated seventh-grade math class. Part II.B addresses topics in accelerated Math 7 that are covered in addition to those in the regular

Math 7 sequence. The additional topics in accelerated Math 7 are the same as those covered in regular Math 8.

Scope and Sequence for Math 7

The focus of this book is limited to integer operations, algebraic operations/reasoning, and ratios and proportions. The topics of geometry, probability, and statistics are not included in this book.

The lessons included in this book are bolded. The numbers indicate the chapter number for the lessons in this part. Lessons contain a brief summary of the content of topics that occur prior to the lesson, and which pertain to the lesson, but are omitted from the book.

Part II.A.

Integers

1. Negative Numbers: Gains and Losses

2. Negative Numbers: Adding Integers on a Number Line

3. Negative Numbers: Subtracting Integers

4. Negative Numbers: Multiplying Integers

Negative Numbers: Dividing Integers

Rational Numbers

Rational Numbers (General Definition, Ordering)

Adding Rational Numbers

Subtracting Rational Numbers

Multiplying and Dividing Rational Numbers

Expressions and Equations

Algebraic Expressions

5. Algebraic Expressions: Translating from English to Algebra

6. Order of Operations

Simplifying Expressions

Associative and Commutative Rules

Distributive Property

7. Solving One-Step Equations with Addition and Subtraction

8. Solving Equations with Rational Coefficients

Two-Step Equations

Two-Step Equations with Fractions

9. Solving Word Problems with Equations

Inequalities

Writing and Graphing Inequalities

Solving Inequalities Using Addition or Subtraction

10. Solving Inequalities with Multiplication and Division

Solving Two-Step Inequalities

Ratios and Proportions

Ratios and Rates

11. Ratios and Tape Diagrams

12. Solving Ratios Algebraically

Unit Rates

Writing Proportions

13. Solving Proportional Relationships

Slope

Direction Variation

14. Other Aspects of Proportions

Graphing Proportional Relationships

Percents

15. Equations with Variables on Both Sides

16. Percents and the Percent Equation

17. Solving Percent Problems Using Proportions

Percent of Change

18. Discounts, Mark-Ups, Taxes, and Tips

19. Unitary Method for Discounts and Mark-Ups

20. Finding Original Amounts

Simple Interest

Constructions and Scale Drawings

Adjacent and Vertical Angles

Complementary and Supplementary Angles

Triangles

Quadrilaterals

Scale Drawing

Area

Circles and Circumference

Perimeters of Composite Figures

Areas of Circles

Areas of Composite Figures

Surface Area and Volume

Surface Areas of Prisms

Surface Areas of Pyramids

Volumes of Prisms

Volumes of Pyramids

Volumes and Surface Area of Composite Figures

Part II.B: Accelerated Seventh Grade and Math 8

Graphing and Writing Linear Equations

21. Linear Equations and Functions

22. Representing Linear Functions

Graphing Linear Functions

23. Slope and Constant Rate of Change

Graphing Proportional Equations and Direct Variation

24. Slope-Intercept Form of Linear Equations, and Graphing

25. Systems of Equations: Solving by Graphing

26. Systems of Equations: Solving by Elimination

Systems of Equations: Word Problems

Writing Equations in Slope-Intercept Form

Transformations

Translations

Reflections

Rotations

Dilations

Scale Drawings

Real Numbers and the Pythagorean Theorem

Volumes of Cylinders, Cones, and Spheres

Composite Volumes and Surface Areas

Probability

Experimental and Theoretical Probability

Compound Events

Independent and Dependent Events

Statistics: Samples and Populations

Statistics: Comparing Populations

PART II.A

1. NEGATIVE NUMBERS: GAINS AND LOSSES

In my Math 7 classes, I start the beginning of the school year with the unit on integers, which includes operating with negative numbers. It is not uncommon for students to understand and carry out a procedure perfectly the day it is introduced, only to totally forget it the next day. It is therefore important to refresh the procedures for operating with negative numbers throughout the year to ensure what has been mastered stays that way.

Some aspects of negative numbers are abstract to students. Procedures may not make logical sense to them. The mathematician John Von Neumann once said, "In mathematics you don't understand things. You just get used to them." Like many topics in mathematics, after experience and practice with a procedure and concept, what was once alien becomes familiar. The familiarity eventually allows students to accept the concept as reasonable—they may even wonder why they ever found it confusing.

Some students may have learned the operations with negative integers in sixth grade, depending on what textbook they used. For those students, this unit will be a review. I start with what students may know about negative numbers. I use examples such as temperature (degrees below zero), or depth (feet underground or under water). We cover the general concepts of gains and losses, and their representation. A gain of \$10 can be represented as +\$10 or \$10. A loss is represented as –\$10. A descent of 50 feet can be represented by the number –50.

I review the number line. The main points: 1) negative numbers are numbers that are less than, or to the left of, zero; positive numbers are to the right of zero; and 2) the bigger the number, the farther it is to the right; the smaller the number, the further it is to the left.

I also review absolute value. I will ask if –4 and 4 are the same distance from 0 on a number line. Absolute value is defined in early grades as distance from zero on the number line—and since distance is always positive, so is absolute value. Whether a number is to the left or right of zero, the distance from zero is always positive.

WARM-UPS

1. Divide. $20 \div 8$. (Express as a decimal.) *Answer*:2.5.
2. Find the value of x.$x \div 10 = 2$. *Answer*: 20.
3. A car travels at a constant speed of 55 mph. How many miles has it traveled in 3 hours? *Answer*: $55 \times 3 = 165$ miles.
4. Find the average of the numbers: 47, 48, 64. *Answer*: 53.
5. Solve. $876 \times 45 \times 98 \times 324 \times 0 \times 34 \times 1{,}201$. *Answer*: 0.

These problems help me to evaluate students' proficiency in and understanding of previously learned material. For Problem 5, students who see the zero in the middle of all the other numbers will get the answer quickly. This problem emphasizes the need to read the entire problem before delving into it. The only prompt or hint I have given to students multiplying all the numbers is: "Have you looked at all the numbers in the problem?" That usually does the trick.

GAINS AND LOSSES

This lesson introduces students to the concept of adding negative integers. The approach for this introductory lesson comes from JUMP Math, and it uses the concept of gains and losses. I find it to be an effective way to introduce students to adding negative integers without them realizing that

that's what they're doing. After the first few minutes of working with the problems, it is amazing to see them doing intuitively what they will be doing when learning the formal rules for adding negative integers.

I start by asking the question, "How do we represent a loss of 5 pounds?" Hearing –5, I continue. "Does a person who loses 5 pounds and then gains 5 pounds end up weighing more, less, or the same as their starting weight?" Students will generally agree that they will end up with the starting weight.

"What about a gain of $6 and a loss of $6?" Students will answer that the person will break even.

Writing +7 –4 on the board, and explaining that the numbers represent money, I ask, "Let's say the numbers indicate a gain and loss of money. If there is a net gain, we call it a good day; a net loss is a bad day. Which is it?" They see immediately that it was a good day.

"How good was it? How much was gained?" They are quick to tell me $3.

"What about –6+2? They tell me it is a loss of $4.

"Good. You have more of a loss than a gain. That means you have more negative numbers than positives. What you did in your heads was to solve it as an everyday subtraction problem with the signs reversed: 6 – 2. Since it is a loss, we're going to write it as –4."

I point out that if the first number is positive, as in +7-8, we don't have to write the positive sign. Additional examples help get students used to this, although it may take longer than you would like before there are no blank stares when you omit the "+" sign from the first number.

Examples. I want you to tell me if it is a net gain or loss, and by how much. If it's a loss I want you to indicate it with a negative sign.

1. –5, 3. *Answer*: *Net loss*: –2.
2. 2 + 2. *Answer*: *Net gain*: 4.
3. –7, –7. *Answer*: *Net loss*: –14.

A common error is to interpret two losses of the same number as zero (e.g., – 7 – 7 is mistakenly thought of as 7 – 7). This mistake will come up repeatedly, and the remedy that I have taken is to remind students of what it represents in terms of two monetary losses: "I lost $7, and I lost $7 more; how much did I lose in all?"

Adding More than Two Gains and Losses

"Let's up the ante a bit. Do I have a net gain or loss?" I write on the board: 3–4–5.

Students might stare at this with no answers forthcoming. "Just do it with the first two numbers. Is there a gain or loss with the first two numbers?

Response: –1.

We're left with –1–5, which students will know is a loss represented as –6. "An easier way is to add the total gains, then add the total losses. In this way, we get +3 –9 for an overall loss of 6, which we write as –6."

With this advice I then give students the next problem: 2–5–4+8–3. They do this in their notebooks. Adding the gains, they get 2 + 8, or 10. The losses are –5–4–3. Total gains equal 10, and total losses are 12: 10–12, for a total loss of 2, or –2.

Examples

1. –6,–5,10. *Answer*: –11,10; *Net loss*: –1.
2. –2,100. *Answer*: *Net gain*: 98
3. 1,–2,3,–6,4. *Answer*: 8,–8; *no change.*

Homework. For homework, I assign problems from JUMP Math (see Figure 1-1). Problems can also easily be constructed as a worksheet.

NS7-3 Adding Gains and Losses

1. Was it a good day (+) or a bad day (−)?

a) +5 − 3 __+__ b) +3 − 5 ____ c) −4 + 3 ____ d) +6 − 9 ____

e) −3 + 2 ____ f) −6 + 7 ____ g) +7 − 2 ____ h) +7 − 10 ____

2. How much was gained or lost overall?

a) +6 − 5 ____ b) −5 + 3 ____ c) +5 − 5 ____ d) −6 + 6 ____

e) +4 + 2 ____ f) −1 − 3 ____ g) +6 − 2 ____ h) +3 − 3 ____

To add many gains and losses:

Step 1: Circle the gains.

Step 2: Add the gains and losses separately.

Example: (+3) − 4 − 5 (+6) (+2) − 5 (+1)
= +12 − 14 = −2

3. Circle the gains. How much was gained or lost overall?

a) (+3) − 2 − 8 (+4)
= +7 − 10
= −3

b) +6 + 5 − 3 − 7 + 6 − 4

c) −5 − 6 + 9 − 8 + 10

d) −9 + 7 − 2 + 1 + 1

e) −1 − 1 − 1 − 1

f) +1 − 2 + 3 − 4 + 5 − 6 + 7 − 8

Figure 1-1: Gain/loss problems from JUMP Math from AP workbook 7.1, Common Core edition, 2015; Toronto (printed with permission).

2. NEGATIVE NUMBERS: ADDING INTEGERS ON A NUMBER LINE

This lesson now extends the gains and losses method to adding numbers on the number line. Students may think that the number line method is different from the gains and losses method, and that each will be used to solve certain types of problems that differ from one another. It is therefore important to tell students that the number line method is a way to look at what was happening in the previous lesson when we worked with gains and losses.

This lesson also presents the rules for adding integers, which break down to three cases: 1) adding two positive numbers, 2) adding two negative numbers, and 3) adding a negative and positive number. In general, I reframe problems in the gain/loss format whose rules are the same as the formal ones, and which most students grasp intuitively. My goal is to capitalize on their intuitive notions.

Warm-Ups

1. John rides his bike 3 miles to the east; he then turns around and rides the same road for 2 miles to the west. How far and in what direction is he from where he started? (Draw a picture like a number line to help you.) *Answer: He will be 1 mile west from his starting point.*
2. He rides his bike again the next day: 4 miles to the east and 6 miles to the west. How far and in what direction is he from where he started? *Answer: He will be 2 miles east from his starting point.*

3. A football team gains 8 yards on the first play. They lose 10 yards on the second play. Do they have an overall gain or loss and by how much? *Answer: They have a loss of 2 yards, or –2.*
4. Gain or loss and by how much? 5–15. *Answer*: *Loss of* 10: –10.
5. Gain or loss and by how much? –3–4. *Answer*: *Loss of* 7:–7.

Problem 1 is best solved by drawing a diagram. It may be illustrated by drawing the diagram on the board. Students can then use that technique to solve Problem 2. Problem 3 is a gain/loss problem. Prompts may include "Good day, or bad day?" and "Are they ahead of where they started or behind? How much?" Problems 4 and 5 are gain/loss problems like students had on their homework.

Using the Number Line

I use Problem 3 of the Warm-Ups to start off the lesson. "When you did Problem 3, and you put it in 'good day/bad day' format, you were able to solve it fairly quickly. It's just what we did yesterday. We wrote it as 8 – 10. I'm going to write it differently now, but it's the same problem." I write on the board: (+8) + (–10)

The purpose of writing it this way is to emphasize that we are *adding a loss* of 10 yards. I draw a number line on the board and ask students to do the same in their notebooks (see Figure 2-1).

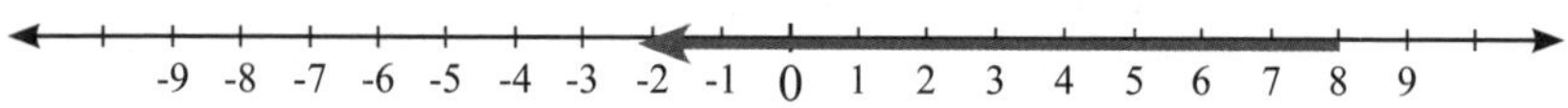

Figure 2-1: Number line representation

"Here are the rules for showing this problem on a number line. We mark the first number on the number line. Now, if we were adding a positive 10, we would draw an arrow starting at 8 and go 10 units to the right. But we're not. We're adding –10. So, what direction do you think the arrow is going to go?"

Response: Left.

"Draw an arrow 10 units left of 8. Tell me where it ends."

Response (many students): –2.

"You'll notice this is the same answer you got when you did the Warm-Ups using the good day/bad day method."

Examples

Students work on the examples in their notebooks as I go around, offering guidance and help as necessary.

1. (+4)+(–4). *Answer*: 0.
2. (+7)+(–12). *Answer*: –5.
3. (–5)+ (–5). *Answer*: –10.
4. (–8)+(+2). *Answer*: –6.

After it appears that students have the knack of doing problems on the number line, I then pick one of the examples, say (+7) + (–12). "I'm going to write this problem like we did yesterday. It's the exact same thing as +7–12, or 7–12. I'm going to write a pattern on the board to help you with this."

I write on the board: +(+) = +; +(–) = –.

"Let's try some examples. What's +(+3) + (–7) without parentheses?

I should hear 3 – 7.

"Try these without parentheses."

1. (–7)+(–8). *Answer*: –7–8.
2. (–9)+(+8). *Answer*: –9+8.

I do other problems until I hear more correct answers than incorrect one.

Summary of the Rules for Adding Negative Numbers

"You aren't going to use a number line every time you want to solve a problem, like –100+57. You already know how to do these from the good day/bad day technique. But here is the formal rule for adding integers."

1. The sum of two positive numbers is positive.

"For example, I gained 3 pounds last week and 2 pounds this week. Total gain is 3 + 2, or 5."

2. THE SUM OF TWO NEGATIVE NUMBERS IS NEGATIVE.

"Who can give me an example of this?"

I may have to hint that losing money or losing weight are good ones to use. I will usually have a few students come up with examples such as "I lost $5 and then lost $2 for a total loss of $7."

3. ADDING INTEGERS WITH DIFFERENT SIGNS

Since students have been working with gains and losses to add negative numbers, I state the rule informally.

1. Determine whether the sum represents a gain or a loss.

2. Find the difference between positive values of the numbers.

3. If it's a loss, give the answer a negative sign.

4. If it's a gain, it will have no sign, since no sign means positive.

"Somebody tell me the answer to this: Our team gained 4 yards and then lost 6 yards. Are we ahead or behind and by how much?"

Students tell me it is a loss of 2 yards.

"I can write this problem as 4 – 6. The amount of loss is greater than the amount gained, so there is an overall loss. Tell me how you calculated the loss."

I will call on someone who I am fairly certain knows the answer in order to keep things flowing. They will say that they calculated it as 6 – 4.

"How do I write the loss of 2?"

Response: –2.

FORMAL RULE

In addition to the informal summary of adding numbers with different signs, I provide students with the formal rule for them to paste into their notebooks.

Subtract the lesser absolute value from the greater absolute value. Then, use the sign of the integer with the greater absolute value.

Example: –5 + 3. The absolute values of the two numbers are 5 and 3. The integer with the greater absolute value is –5; since the sign is negative, the answer is – (5–3)= –2.

The formal rule will make more sense at a later time, after they work with these types of problems more. For now, they work with the informal rule.

Common Error

The error of thinking of –5 – 5 as 0 will persist. Since students have been working with number lines, I show those who make this error what –5 –5 looks like on the number line, and then show (+5) + (–5) and (–5) + (5) on the number line as well.

Homework

Problems are the same as the previous lesson, except a few require use of the number line. The first few problems are written with parentheses and signs. The remaining are written without parentheses as they appeared in the previous lesson. I work with students as they start on the homework problems, observing which students are having the most difficulty and making sure they understand how to do the problems.

3. NEGATIVE NUMBERS: SUBTRACTING INTEGERS

This lesson provides clarification and, for some, a revelation, that subtraction of two numbers is the addition of an additive inverse. In formal mathematical terms, $x - y$ is defined as $x + (-y)$, where $-y$ is the additive inverse (or opposite) of y. Some students become confused at this wording and ask, "So subtraction is the same as adding?" This gets somewhat cleared up when they see that 7 – 4 is the same as 7 + (–4), as is demonstrated at the start of the lesson. (My co-author eliminates confusion by referring to both addition and subtraction of integers as "combining integers.")

Up to this point, this lesson is a clarification of what has come before. One important case that hasn't yet been explored is subtraction of a negative integer, such as 5–. Subtraction of a negative number will be the new procedure.

Warm-Ups

1. The temperature yesterday was –5°F. Today's temperature is 25 degrees higher. What is today's temperature? *Answer: It takes an increase of 5 degrees to reach zero. Another 20 degrees makes a total incease of 25 degrees. Twenty degrees more than zero is today's temperature: 20 degrees.*
2. Which is greater? $|-5 + (-7)|$ or $|-5 + 7|$? Show your work. Answer: $|-5 + (-7)| = |-12| = 12$; $|-5 + 7| = |2| = 2$. $|-5 + (-7)|$ *is greater.*
3. Your team lost 5 yards on a play and received a 15-yard penalty. What is their net gain or loss? Show your work. *Answer*: –5–20 = –25; *or* –5 + (–20) = –25

4. –6 + 6 = ? *Answer*: 0.
5. Peter, John, and Dan shared $1,458 equally. Peter used part of his share to buy a bicycle and had $139 left. What was the cost of the bicycle?

   Answer: 1458 ÷ 3 = 486.486 – 139 = $347.

Problem 1 is solved intuitively, and most students get the correct answer. Problem 2 requires knowledge of combining integers as well as absolute value. Problem 3 is the addition of two negative numbers. The context of the problem, however, makes it obvious that the answer will be the sum of the negative numbers. Problem 4 is a reminder that the addition of opposites equal zero. Problem 5 is not related to negative numbers but requires students to solve the problem in stages. I watch for students who might write the equations as a one run-on; that is, 1458 ÷ 3 = 486 – 139 = 347. This is equivalent to saying that 1458 ÷ 3 = 347, which is wrong.

Adding the Opposite

I start this lesson by asking the class to find 7 + (–4). After they do, I say, "Now tell me what 7 –4 equals."

This will seem like something new to them even though they have already been doing it with the good day/bad day exercises. We also discussed in the previous lesson that 7 + (–4) can be expressed as 7 – 4. To hammer home the point that this is the same problem, I say so.

"Whenever you are adding negative integers, you are subtracting. Subtraction of an integer is the same as adding the opposite of that integer. The problem 7 – 3 is the same as adding the opposite of 3, which is 7 + (–3). You've been doing this for the past few days."

What Subtraction Represents and Subtracting a Negative Integer

As an introduction to subtraction of a negative integer, I talk about what subtraction represents.

"A subtraction problem can represent different things," I begin. "For example, a problem like 10 – 4 can represent how much money is left over from $10 after spending $4. If, however, the question were 'The temperature was 4 degrees and now it is 10 degrees; by how much did it change?', the answer is still 6, but the numbers represent different things. The first problem answered 'how much is left?' The second answered 'what is the difference?' In both cases, the problem is represented as 10 – 4."

"Now suppose we have this problem." I write on the board:

10–(–4)To keep things straightforward, I limit the examples of what this may mean to the second model—that is, "What is the difference?" rather than what a loss of –4 represents since that will confuse more than enlighten.[1]

"Let's say it was –4 degrees yesterday, and today it's 10 degrees. What is the increase in temperature? We had a problem like that on the Warm-Up."

Drawing a number line (vertical, to look like a thermometer), I plot –4 and 10. "What is the increase in temperature needed to get from – 4 to 0?" The students tell me 4 degrees. "That makes sense, because – 4 + 4 = 0. The sum of opposite numbers is zero."

Transferring this to the thermometer number line, it is apparent that 0 to 10 degrees is an increase of 10 degrees. "We had an increase of 4 to get to 0, and now another 10 degrees to get to 10; what's the total?"

Some of the lights come on in the students.

"It isn't always practical to use number lines to solve problems. If you had 120 – (–24), you'd want a more compact way. We can use the 'add the opposite' rule. The problem 10 – (– 4) becomes 10 plus the opposite of –4, or 10 + 4, which is 14."

1 An example of a problem that asks what's left after a loss of –4 would be: A person has $10 in their bank account after $4 has been deducted in error. The bank corrects this error by removing the debit of $4. This is done by subtracting the loss of $4: 10 – (–4), which then becomes 10 + 4, or $14. Even adults may find it confusing that cancelling a debt can be represented by subtraction of a negative number. It is therefore highly likely that seventh-graders will find the concept difficult. Since the goal is for students to subtract negative numbers, it is far easier to explain the procedure using the "find the difference" model discussed above, rather than by the "find what's left" model.

Another example using depth underwater illustrates subtracting a negative number as well. I scaffold the problem by first starting with all positive numbers: "A bird was 5 feet in the air and flew up to 10 feet. What was the distance upward that it flew?" Students will easily see that it is "new height minus original height," or 10 – 5.

"Let's say that a bird dives underwater 5 feet to catch a fish. How do we indicate its depth underwater?"

Response: – 5.

"It then flies upward to a height of 10 feet above the water. What was the upward distance that it flew? I'm finding the difference between what two numbers?"

Not all students will grasp this immediately. Some will tell me 10 – (–5). "And what does that become? What's the opposite of – 5?" I hear +5, and they add the numbers.

I also want to show problems where there is a decrease. For example: "The temperature today is -5 degrees; yesterday it was 6 degrees. What is the change in temperature?" The model for the problem is "Today's temperature – yesterday's temperature," which I write on the board so students see the pattern. The problem is written -5 - 6, which equals -11, or a decrease of 11 degrees.

"What if I said it's –5 today, and it will decrease 6 degrees tonight. What will the temperature be?" Students may struggle with this; to give them a hint via a type of problem they've solved before. I will prompt them with, "Suppose I had said it's 50 degrees today, and it will decrease 6 degrees; what will the temperature be?" Most will see what to do and apply it to the problem: -5-6=-11. I now give them four or five problems (taken from the homework to give them a start on it).

Homework

Most of the problems require subtracting a negative, but others, like the last one, are adding two negative integers or adding one negative and one positive number. Some problems will be word problems, but most are numerical computations.

Giving It a Rest

Mastering the operations with negative numbers will be confusing for some students at first. Although these lessons continually refer to and reinforce the "gain/loss" procedure, some students may become overwhelmed by new information as well as the mathematical way of stating things. In particular, they are now learning to think of subtraction as the addition of an opposite number.

It is advisable to give students time to work with the newly learned procedures and to provide opportunities for practice.

4. NEGATIVE NUMBERS: MULTIPLYING INTEGERS

Before moving on from the operations of addition and subtraction to multiplication and division, I provide a wrap-up of what students have learned. This includes the concepts that students have learned for addition and subtraction in the lower grades, where all answers are restricted to positive integers. For example, for the problem $6 - 2 = 4$, to check if it is correct, we add 2 to 4 to make 6. It answers the question "What number added to 2 equals 6?" Similarly, for the addition problem $5 + 2 = 7$, we subtract 2 from 7 to get 5. These rules apply as well to negative integers. The subtraction problem $3 - 6 = -3$ answers the question "What number added to -6 equals 3?"

The multiplication of negative integers can be a confusing topic for students—particularly the rule that the product of two negative numbers equals a positive number. The main problem that students have is understanding what multiplication by a negative number represents.

Providing examples of situations modeled by multiplication with negative numbers is effective in helping students understand the rules. After establishing the rules via the examples, I admonish students that the examples only suggest that these rules are true. The mathematical proof of the rules—generally shown in algebra—relies on the distributive rule.

It is fairly common for students to confuse the rules for multiplication with the rules for addition. For example, upon being given the problem $-7 -7$, students think that the two negatives make a positive 14. The usual reason: "I thought two negatives make a positive." To help alleviate the confusion, I mix multiplication and addition problems in homework assignments so that students are not just mastering the latest rule learned and forgetting the others.

Warm-Ups

1. John loses 10 pounds every month. How much does he lose in 3 months? *Answer*: $10 \times 3 = 30$.
2. Sarah loses 5 pounds every month. How much more did she weigh 4 months ago? *Answer*: $5 \times 4 = 20$ lb.
3. $8 - 9 + 1 = ?$ *Answer*: $-1 + 1 = 0$.
4. $-7 - 50 = ?$ Check the answer and show your work.

   Answer: $-7 - 50 = -57$; Check: $-57 + 50 = -7$.
5. $-1 + 2 - 3 + 4 - 5 + 6 = ?$

   Answer $-1 - 3 - 5 + 2 + 4 + 6 = -9 + 12 = -3$.

Problems 1 and 2 set the stage for the lesson on multiplying negative numbers, though the problems are stated in terms of positive numbers. Problems 3–5 review the addition of negative numbers.

Multiplying Two Numbers

"How would you write the number representing a gain of 4 pounds? Is it positive or negative?" I ask.

Response: Positive.

"What about a loss of 14 pounds?"

Response: −14.

Now, I change things slightly. "Five minutes from now." There is some hesitation.

Response: 5.

"Ten minutes ago." Again, there is hesitation.

Response: –10.

"With that as introduction and background, let's move on to see what's going on when we multiply with negative numbers."

Rules for Multiplying Negative Numbers

I start with positive times positive, then move on to positive times negative, and vice versa. Problem 1 of the Warm-Ups provides a preview of the type of reasoning used in multiplying negative numbers. All numbers are stated as positive, however, so that in Problem 1 we have "a loss of 30 pounds" rather than a change of -30 pounds.

Problem 1 could have been written (−10) + (−10) + (−10), which would let us say what 3 × (−10) means. But giving meaning to (−3) ×10 using the same type of situation as in Problem 1 would be difficult, if not impossible. I therefore use a different type of example based on one used in Dolciani's 1962 version of *Modern Algebra: Structure and Method*, and a similar one in JUMP Math.

"Let's say we have a water tank. Water flows into it at the rate of 10 gallons per minute. Is 10 gallons positive or negative in this case?"

Response: Positive.

"What about 3 minutes?"

Response: Positive.

"At that rate, how much more water will there be in the tank after 3 minutes? That is, what's the change in the amount of water in the tank?" The students tell me it is 30.

This suggests the rule they all know:

3 × 10 = 30: Positive number × positive number gives positive number.

"Now, how many gallons less was in the tank 3 minutes ago, if water is flowing into the tank at 3 gallons per minute?" Students will shout out "30."

"And how do we represent '3 minutes ago' ?" I will hear someone, usually hesitant, say –3.

"Yes, correct. And how do we represent 30 gallons less?"

Hearing -30, I write:

(–3) × 10 = -30: Negative number × positive number gives negative number.

"Now, assume water is flowing *out* of the tank at the rate of 10 gallons per minute. How do I write that?"

Catching on to the pattern, students respond -10.

"What is the change after 3 minutes?"

Recognizing that 3 minutes is a positive value, students tell me it's a loss of 30 gallons.

"Which we can write as what? Positive or negative?" They tell me negative, and I write:

3 × (−10) = −30: Positive number × negative number gives negative number.

The final example is a negative number times a negative number. Before I present the example, I ask if anyone knows whether the product will be negative or positive. "Let's find out. Now, assume water is flowing out of the tank at the rate of 10 gallons per minute. How do I write that?" They tell me −10.

"I want to know if 3 minutes ago there was more or less water in the tank, and by how much." At this point, most students will know the tank held 30 gallons more water.

"And how do I represent 3 minutes ago?" They tell me −3. I now write:

(−3) × (−10) = 30: Negative number × negative number gives positive number.

There will undoubtedly be some students who will not understand how the examples work. I tell these students that it will become clearer the more they work with such problems, but that for now they should just follow the rules and "trust the math." I write the rules on the board:

1. The product of two integers with different signs is negative.

2. The product of two integers with the same sign is positive.

EXAMPLES

1. (−5) × 7. *Answer*: −35.
2. (−2) × (−4). *Answer*: 8.
3. 6 × (−1). *Answer*: −6.

4. (6)(−1). *Answer*: −6. I explain that we can use parentheses to represent multiplication.
5. (7)(−1) + 7. *Answer*: 0. Students may need remininding to follow the order of operations; multiplication comes first and then addition.

Before I set students loose on homework, I put up a problem and ask how I would solve it: (−2)(3)(−5). Such problems are broken down and solved by multiplying two numbers at a time. The above problem then becomes (−6)(−5), which is 30. A problem like (−5)(2)(−2)(4)(−1) becomes (−10)(−8)(−1), which is –80.

Homework

Problems are a continuation of the above, with some word problems mixed in. Also, some addition and subtraction problems involving negative numbers are included, to keep the rules for each set of operations fresh and to prevent conflation of the rules (e.g., assuming incorrectly that the sum of two negative numbers will be positive), as discussed earlier.

5. ALGEBRAIC EXPRESSIONS: TRANSLATING FROM ENGLISH TO ALGEBRA

PRIOR CONTENT

The unit on negative integers concludes with a lesson on division of negative numbers. Division of negative numbers follows as an application of the rules for multiplying negative numbers. The rules follow the same pattern as multiplication:

1. The quotient of two integers with like signs is positive.

2. The quotient of two integers with unlike signs is negative.

Since students know that $2 \times (3) = -6$, they see that they can divide the product, -6, by either 2 or -3 to get the other factor. That is, $-6 \div (-3) = 2$ because $2 \times (-3) = -6$ and $-6 \div 2 = -3$. Some students see the inverse relationship and operate from that. Most students, however, use the rules regarding like and unlike signs as guidance.

We then move on to rational numbers: what they are, and how do we use operations with them. For all practical purposes, rational numbers are treated as fractions and decimals (terminating and repeating), although students learn that whole numbers are rational numbers as well. Negative fractions and decimals are included. I teach ordering of rational numbers on the number line so that students can recognize that, for example, 0.09 is less than 0.2, and that –0.9 is greater than –0.95.

The unit on rational numbers extends students' knowledge of working with negative integers to fractions and decimals (e.g., $-\frac{2}{7} + \frac{3}{5}$ and 0.23-0.5). These require students to evaluate the larger absolute values of the fractions or decimals involved. Multiplication and division of rational numbers is then covered with typical problems, including $-\frac{3}{4} \times \frac{5}{6}$ and $\frac{2}{3} \div (-\frac{3}{4})$.

From rational numbers, it's on to the foundations of algebraic expressions and one- and two-step equations.

Prior to the lesson discussed here, students will have learned what variables are and that unknown quantities are represented with a letter. "Three more than some number" is written $3 + x$, for example. Similarly "a number added to another" can be written as $x + y$. Students have also learned the basics of addition, multiplication, and division with variables. For example, $x + x + x$ is represented as $3x$, meaning 3 multiplied by the variable x. Division in algebra is represented by the fraction bar. The divide sign, ÷, is not used, particularly when using letters.

Lastly, students learn to evaluate algebraic expressions by plugging in given values to a specific expression. For example, $\frac{5}{x}$ when $x = 2$ is $\frac{5}{2}$. $5ab$, where $a = -1$ and $b = 2$, is equal to -10.

In this lesson, students learn basic rules for expressing English expressions in algebraic terms. For example, the expression "the cost of some number of games of bowling at $4 per game" can be expressed as $4x$. "Four less than two times a number" can be expressed as $2x - 4$.

WARM-UPS

1. Write the numerical expression for "5 more than 10." *Answer:* 10 + 5 or 5 + 10.
2. Write the numerical expressions for "5 less than 10." *Answer*: 10 − 5.
3. Write as an algebraic expression: "seven times x." *Answer*: $7x$.
4. $\frac{2-5}{6-9} = ?$ *Answer*: $\frac{-3}{-3} = 1$.
5. $-\frac{5}{8} + \frac{2}{3} = ?$ *Answer*: $\frac{-15+16}{24} = \frac{1}{24}$

Problems 1 and 2 segue to today's lesson by patterning how English expressions are written in symbols. Problem 3 was covered in a previous lesson and is covered further in this one. Problems 4 and 5 combine knowledge of working with negative integers and fractions.

Defining the Variable

"A variable is a letter that represents an unknown quantity," I say. "We saw this in Problem 3 of the Warm-Up: "Seven times " "s written as $7x$. What if I had said, 'Seven times some number'?"

Someone will know the answer: $7x$, which is the same as the problem I just gave.

"Suppose I say, 'Five more than some number.' What is this in algebraic form?"

Response: $5 + x$ or $x + 5$.

Examples. I ask students to write answers in their notebooks.

1. Five times a number. *Answer*: $5x$.

2. Two more than three times a number. *Answer*: $2 + 3x$.

Students may hesitate on this last problem; a prompt may be: "What is three times a number?" "How do you show two more than that quantity?"

3. Three divided by some number. *Answer*: $\frac{3}{x}$.

Some students may write it as $3 \div x$, which is correct, but I admonish them to use the fraction bar instead of the divide sign.

4. Two miles more than an athlete ran. *Answer*: $2 + x$.

For this problem, I tell students not to put the unit "miles" in the answer. A prompt may be: "Do we know how far the athlete ran?"

More Complex Expressions

"Let's look at some more expressions that are a bit more complex. I'm going to have you translate this expression, but I'm betting that there will be more than one person who gets this one wrong. The expression is 'five less than some number."

Usually several people will say $5 - x$ rather than $x - 5$. For those who get it right, I ask them to explain why they wrote it that way. I want to hear something along the lines that five is being subtracted *from* a number.

"If a statement is confusing, and you're not sure how to translate it, write it using numbers. In our Warm-Ups, we had '5 less than 10,' and you had no problem writing that as $10 - 5$. This allows you to see what is being subtracted from what. Substituting x in for 5, you get the correct expression."

(At this point, I hold off on explaining that the phrase "five is less than some number" is different from "five less than some number." The former is $5 < x$. I teach this later in the unit on inequalities.)

"How would I write the cost of some number of games of bowling at \$4 per game?" Some students will tell me $4x$, whereas others may say \$$4x$.

"When we write an algebraic expression, we leave off the units. Now, shall we make it a bit more complicated?"

Someone will inevitably say, "How about no?" which cues me to then give the problem. "How would you write the cost of bowling some number of games at \$4 per game if you also have to pay \$5 for renting bowling shoes?" I ask students to write the answer in their notebooks.

A possible prompt may include "Are you adding something to the cost of four games?" The answer I'm looking for is $5 + 4x$ or $4x + 5$.

Examples. Students should provide the expression in terms of the variable. (These are taken from Brown et al., 1988.)

1. The Tigers had twice as many hits as the Yanks. If x = the number of hits by the Yanks, then ____ = number of hits by the Tigers. *Answer*: $2x$.
2. The length of a rectangle is four times the width. If x = the width, then ____ = the length. *Answer*: $4x$.
3. Mac is x years old. How old will he be next year? *Answer*:$x+1$.

   For this problem, I remind students that they should express Mac's age in terms of x.
4. Trish is t years old. How old was she 7 years ago? *Answer* $t - 7$.

5. Karen will be m years old next year. How old is she this year? *Answer*: $m - 1$.

Possible prompt: "Use numbers. If she'll be 10 next year, how would you figure out her age this year?"

HOMEWORK

Problems include translations from English to algebra, with the story-type problems included, as shown above. Also included are expressions that will be evaluated given values for the variables.

6. ORDER OF OPERATIONS

An expression such as $6 + 4 \times 3$ may be interpreted as 10×3, or as $6 + 12$. This is where "order of operations" comes in—an agreed-upon hierarchy that eliminates ambiguity. (The lesson on this topic for the sixth grade is also included in this book. See Part I, Chapter 3.)

I recall learning about order of operations in my algebra class. What stuck with me after learning the rules was Mr. Dombey's final advice: "When in doubt, use parentheses." This simple advice is how most ambiguities in math are resolved. The need to interpret an expression that does not have grouping symbols, such as parentheses or a fraction bar, doesn't come up very often.

Order of operations is not my favorite topic. Algebraic symbols take order of operations into account by virtue of the rules that govern them. For example, to find the value of $ab + 6$, where $a = 2$ and $b = 3$, you would multiply the a and b values together, since that's what ab means, and then add 6 to the product to obtain 12. You would not multiply a by the value of $b + 6$, since that's not what the algebraic expression represents. But if you did want it to be evaluated that way, you would write it as $a(b + 6)$.

All that said, students are expected to know the rules, and the questions do show up on standardized tests. In the end, however, Mr. Dombey's advice has prevailed.

Warm-Ups

1. Translate into symbols. Five more than three times a number. Answer: $5 + 3n$.
2. Translate into symbols. Six less than two times a number. *Answer*: $2x - 6$.

3. Fill in the blank with <. >, or = . $-\frac{17}{24}$____ $-\frac{11}{12}$. *Answer*: $-\frac{17}{24} > -\frac{22}{24}$.
4. How many $\frac{2}{3}$ oz servings of yogurt are in a $\frac{3}{4}$ cup?

   Answer: $\frac{3}{4} \div \frac{2}{3} = \frac{3}{4} \times \frac{3}{2} = \frac{9}{8}$ *or* $1\frac{1}{8}$.
5. A boat rental firm charges a $5 flat fee and $4 for each hour the boat is used.
   a. What does it cost to rent a boat for 5 hours? *Answer*: $5 + 4 \times 5 = \$25$.
   b. Write an expression for how much it costs to rent a boat for n hours.

   Answer: $5 + 4n$.

Problems 1 and 2 review translating English into algebraic symbols. As students work on the problems, I check to see how they are doing Problem 2, since even with warnings students will write it as 6-2x. Problem 3 combines operations with fractions with understanding order on the number line. Problem 4 is a fractional division problem that students tend to have difficulty with. A prompt I give is "How would you solve it if it were 'How many 2 oz servings are in a 6 oz container?' Problem 5 is part of the previous lesson—expressing a situation in algebraic symbols.

Punctuation Marks in Algebra

I start out by asking about some of the Warm-Up problems, which we have just gone over. "For Problem 2, if I said that x equals 4, what would be the value of the expression?"

I call on someone, and once I get the correct answer of 8 (which is usually right away for this type of problem), I ask, "Tell me how you did it."

The student will say they multiplied 2 times 4 and then subtracted 6.

"Aha! And what about the first part of Problem 5? Someone tell me how you did that."

The student will report a similar procedure: multiply 5 times 4 and add 5.

"All of you have learned about order of operations, and you've seen the term PEMDAS, correct? Now whether you were aware of it or not, you were using one of the rules of the order of operations. Tell me what PEMDAS stands for."

Many students seem to know, and at least one person will say, "Please Excuse My Dear Aunt Sally," which stands for: Parentheses, Exponents, Multiplication, Division, Addition and Subtraction.

"We use this order to be clear about how to interpret expressions. The first category, parentheses, can also be called 'groupings' because they include such things as a fraction bar."

I write the following sentence on the board, which also appears in the algebra textbook by Dolciani et al (1962) in its discussion of order of operations: *Paul said the teacher is very intelligent.*

"There is no punctuation in this sentence. We can punctuate this sentence in various ways. I will write one way on the board."

"Paul," said the teacher, "is very intelligent."

"Or, we can write it like this:"

Paul said, "The teacher is very intelligent."

"The punctuation gives the sentence two different meanings. I tend to prefer the second sentence for personal reasons." (Only one student has ever laughed at this little aside.)

"The difference in punctuation lends different meanings to the same sentence. It's the same in math. For the expression $5 + 4 \times 5$, we have some choices for how we want to punctuate the expression. It was evident to all of you that you add 5 to the product of 4 and 5. Where should we place the parentheses to make it clear that's what we want to do? Write it on your mini-whiteboards."

I'll usually see several correct answers: $5 + (4 \times 5)$.

"What if I wanted to punctuate it so that we are to add 5 and 4 and then multiply that sum by 5?"

This one usually takes them a little longer, but they do it: $(5 + 4) \times 5$.

At this point, I digress a bit to introduce a convention in algebraic expressions, which is that we can use parentheses to indicate multiplication; for example, $(5 + 4)5$, or $5(5 + 4)$.

"Now, let's try something. I'll write the mathematical expression, and I want you to state the English equivalent of what various expressions mean. For example: (8 + 3) − 5 can be 'The sum of 8 and 3 minus 5.'"

Examples

1. 2(5 + 3). *Possible answers: "The sum of five and three multiplied by two" or "the product of two and the sum of five and three."*
2. 6 − (2 × 3). *Answer: Six minus the product of two and three.*
3. (30 ÷ 2) + 4. *Possible answer: Thirty divided by two,added to four.*

Order of Operations

"We've seen that parentheses can take the guesswork out of what an expression means. But when parentheses are not used, the order of operations is the agreed-upon rule for interpreting expressions." I write the rules for order of operations on the board:

1. Simplify the expression within each grouping (i.e., parentheses, or fraction bars).

2. Perform multiplication and division in order from left to right.

3 Finally, do the addition and subtraction in order from left to right.

"Now, I'll write some expressions on the board, and I want you to place parentheses so that the expressions now follow the rules for order of operations. Write them on your mini-whiteboards."

Examples

1. 5 + 3 × 2. *Answer*: 5 + (3 × 2).
2. 3 × 4 + 7. *Answer*: (3 × 4) + 7.
3. 30 ÷ 10 ×3. *Answer*: (30 ÷ 10) × 3.
4. 5 − 2 + 4 × 3. *Answer*: (5 − 2) + (4 × 3).

Additional Problems

I give students some additional problems that I take from their homework to get them started and so they know how to do these particular ones.

"I want you to place parentheses to get a particular number for the expression. For example, if we want the expression $2 \times 3 + 5$ to equal 11, the parentheses would be placed as follows: $(2 \times 3) + 5$. If we wanted it to equal 16, it would be $2(3 + 5)$."

1. $4 + 6 \times 2$. *Value* = 16. *Answer*: $4 + (6 \times 2)$.
2. $3 - 2 \times 5$. *Value*: 5. *Answer*: $(3 - 2)5$, or $(3 - 2) \times 5$.
3. $2 + 6 \times 2 - 8$. *Value*: 6. *Answer*: $2 + (6 \times 2) - 8$.
4. $2 + 6 \times 2 - 8$. *Value*: −48. *Answer*: $(2 + 6) \times (2 - 8)$.

Homework

The homework assignment should have a mix of problems that ask students to apply the order of operations, as well as those that ask for parentheses to be placed to obtain a specified value.

7. SOLVING ONE-STEP EQUATIONS WITH ADDITION AND SUBTRACTION

Prior Content

After the lesson on order of operations, students learn how to simplify algebraic expressions. This includes combining like terms; for example, $a + 3b - 2b + 2a$ can be simplified to $3a + b$. Students also learn to multiply terms; for example, $3(5x) = 15x$ and $-2(-3x) = 6x$. Most importantly, they learn that division by zero is not permitted, with an explanation as to why. (This is also discussed in the early grades. See Part I, Chapter 4.)

Following simplification of expressions, students learn the rules of associativity and commutativity, as well as the distribution property. Although these rules have also been introduced in sixth grade so that this is a review (see Part I, Chapter 4). I introduce the algebraic form of the rules. The associativity rules for addition and multiplication are $(a + b) + c = a + (b + c)$, and $(a{\cdot}b){\cdot}c = a{\cdot}(b{\cdot}c)$. Commutativity rules for addition and multiplication are $a + b = b + a$ and $a{\cdot}b = b{\cdot}a$. These two sets of rules taken together allow for simplification of computation (e.g., $4 + 5 + 6 + 3 = 4 + 6 + 5 + 3 = 10 + 8 = 18$). Similarly with algebraic expressions, students learned to combine like terms, no matter where in the expression the like terms occur. The associative and commutative rules make that possible.

After this comes the distributive property: For any numbers, a, b and $c, a(b + c) = ab + ac$. Starting with simple computations such as $3(2 + 3)$, students progress to distributions with variables such as $3(x + 2)$, and w$-2(x - 2)$. This last example proves to be confusing for students. A helpful way to approach distributing a negative number is to show that multiplying by a negative number changes the signs inside the parentheses; for example, $-1(a - 2) = -a + 2$.

The formality of the rules may be abstract, but they are not above the capabilities of seventh-graders. The preface to the book Naïve Set Theory by the late mathematician Paul Halmos (1955) contains a statement about set theory that I find applicable here: "General set theory is pretty trivial stuff really, but if you want to be a mathematician you need some, and here it is; read it, absorb it, and forget it."

Not that I expect all my students to become mathematicians, but I do want to give them what they need, just in case.

TODAY'S AND A SUBSEQUENT LESSON

In today's lesson, I cover solving one-step equations using addition or multiplication. Students have been solving equations by using "number bonds." For each addition fact, such as $a + b = c$, they learn that $c - b = a$ and $c - a = b$. For $n + 5 = 9$, they know that $9 - 5$ will be the answer. Algebra does the same thing by subtracting 5 from each side. Students may find this cumbersome since the arithmetic method they know is easier for these types of equations.

Similarly, with equations in the form $ax = c$; where $a = 1$), students learn to divide each side by a. For equations in the form $\frac{x}{a} = c$, or $\frac{1}{a}x = c$, they learn to multiply both sides by the number in the denominator. Thus, $3x = 18$ becomes: $\frac{3x}{18} = \frac{18}{3}$, leading to $x = 6$; and $\frac{x}{2} = 5$ becomes $\frac{2}{1} \cdot \frac{x}{2}$, leading to $= 5 \cdot 2$. (The lesson on solving equations by multiplication and division is not included in this book.)

WARM-UPS

1. Simplify by distributing. $2(x - 5)$. *Answer*: $2x - 10$.
2. Simplify. $-(-3 - x)$.*Answer*: $-(- 3 - x) = -1(-3 - x) = 3 + x$.
3. Combine like terms.$-x + 5x + 6y - 7y$. *Answer*: $4x - y$.
4. Solve for n. $n - 9 = 5$. *Answer*: $n = 14$.
5. $5 + ? = 0$. *Answer*: -5.

When going through the Warm-Ups, Problems 4 and 5 are germane to today's lesson. Problem 4 of the Warm-Ups is something students have seen since the early grades using arithmetic methods. For Problem 5, students who remember that the sum of opposite numbers is always zero will have no problem with it. There may be others who have forgotten about opposites. This is a key part of the day's lesson, so it is good to go over this.

Addition and Subtraction Properties of Equality

"Today we will learn how to solve equations like Problem 4 of the Warm-Ups, using algebra. It will probably seem like more work than you're used to, but we have to start somewhere. As equations get more complicated, the algebra method is far simpler. At the end of this unit, you will be solving equations like this:"

$$\frac{2}{3}x + \frac{5}{6} = \frac{7}{8}$$

"I think you'll agree it's best to start with something easy and work up." Hearing no disagreement, I write on the board:

$$6{,}000 = 6{,}000$$

"I assume we all agree with what I just wrote. And if I added 500 to each side, will both sides be equal?"

Students agree but wonder where I'm going with this.

$$6{,}000 + 500 = 6{,}000 + 500$$

"We all agree we get 6,500 on each side. What if I wrote the first equation like this?"

$$6000 = 2000 + 4000$$

"If I add 500 to each side, will the two sides still be equal?"

There is general agreement again, but if anyone should doubt it, I'll have them add it up. So far no one has doubted it.

"This is a rule of algebra called the addition property of equality. It's written like this:"

For each number *a*, *b*, and *c*, if $a = b$, *then* $a + c = b + c$.

"It means that if the same number is added to equal numbers, the sums are equal. And the same holds for subtracting the same number from each side."

I write on the board:

$$75 = 50 + 25$$

"Subtract 12 from each side and see if both sides are still equal."

Students get 63 on both sides. (Those who don't generally keep it to themselves.)

"Let's see how we use all this to solve equations."

Solving Equations Using These Properties

I write the following equation on the board: $x + 7 = 25$.

"The name of the game is to get the 7 to the other side so we only have x on the left side. When we do that, we'll have *x* equals something, and we want to know what that something is. This is known as 'isolating the variable.'"

I refer back to Problem 5 of the Warm-Ups about adding opposites, and I ask, "What is the opposite of 7?"

Response: −7.

I write:

$$\begin{aligned} x + 7 &= 25 \\ -7 &= -7 \end{aligned}$$

"We add −7 to both sides of the equation. Since 7 + (−7) is 0, we essentially have $x + 0 = 25 - 7$, or $x = 18$."

"On a test or quiz, if you ask me if the answer is correct, I will not tell you. But you can find out by checking it yourself. Substitute 18 for *x* in the original equation. Does it check?"

Examples. I ask students to do two or three more problems at their desks, writing their work in their notebooks. I tell them to show and check their work.

1. $x + 17 = 19$. *Answer*: $x = 2$.
2. $x + 6 = 2$. *Answer*: $x = -4$.

I remind students that adding a negative number is subtracting and that it's perfectly fine to get a negative number for an answer.

3. $2x - x - 5 = 10$. *Answer*: $x - 5 = 10$; $x = 15$.

There is undoubtedly confusion for this problem, so I will ask, "How do I simplify the left side? Can we combine like terms? What is $2x - x$?" And, also undoubtedly, there will be a few students who keep forgetting that x is the same as $1x$.

Homework

Problems are a continuation of what we have done in class, with later problems requiring simplification on both sides; for example, $3x - 2x + 6 - 4 = 2 - 4$. For these problems, the rules for adding negative integers come into play once more, and some students may need to work on these skills.

8. SOLVING EQUATIONS WITH RATIONAL COEFFICIENTS

After learning about solving one-step equations with addition and subtraction, students learn how to solve one-step equations with multiplication and addition. These problems are taught using two principles: The division property of equality, and the multiplication property of equality. We treat them as two different types of problems, although later when students have had a course in algebra, they will learn that $\frac{a}{b}$ is equivalent to a $\cdot\frac{1}{b}$. Thus, equations in the form $ax = b$ are solved using the multiplication property of equality since both sides are multiplied by $\frac{1}{a}$.

In teaching algebra and Math 8 classes, I make a point of showing students this. For seventh grade, I mention it in terms of what we are actually doing when we solve equations in the form $\frac{x}{m} = b$; that is, $\frac{x}{m} = \frac{1}{m}\cdot x$. For seventh-graders, I mention that $\frac{x}{5} = 20$ can be written as $\frac{1}{5}x = 20$. I mention it because it serves as a connection to solving equations with fractional coefficients, such as $\frac{2}{3}x = 16$.

Warm-Ups

1. $\frac{3}{5} \times \frac{5}{3} = ?$ *Answer*: 1.
2. $0.2 \div 0.05 = ?$ *Answer*: $\frac{20}{5} = 4$.
3. $\frac{1}{7}y = 4$. *Answer*: $y = 28$.
4. $4x - 2x = 238 - 138$. *Answer*: $2x = 100$; $x = 50$.
5. $\frac{2}{3} \div \frac{5}{6} = ?$ Answer: $\frac{2}{3} \times \frac{6}{5} = \frac{4}{5}$.

Problems 1 and 2 provide review of multiplication of fractions and division of decimals. Problem 1 also offers me the opportunity to point out what a reciprocal is, and how multiplication of any number by its reciprocal equals 1. Problem 3 requires both sides to be multiplied by the denominator of the coefficient. Problem 4 requires simplification on both sides.

Fractions as Coefficients

"We learned yesterday how to solve problems like Problem 3 and 4 of the Warm-Ups. I want to take a look at Problem 3 once more."

"I could write the problem as $\frac{y}{7} = 4$. It's the same problem. When I do this, you usually tell me to multiply both sides by 7. But when it is written as $\frac{1}{7}y=4$, some of you tell me to divide both sides by $\frac{1}{7}$. This is actually what we are doing when we multiply both sides by 7. Why, you ask? Let me show you." I write on the board: $y = 4 \div \frac{1}{7}$.

"When we divide by a fraction, what do we do to the fraction we're dividing by and what happens to the sign?"

Students know the "flip and multiply" or "keep-change-flip" mantras, and usually tell me to do so. "What we end up with then is 4×7, right?"

There are some lights that come on here, and maybe even an inkling of why the "flip and multiply" procedure works as it does.

"When the equation is written as $\frac{1}{7}y = 4$, what you are doing is multiplying both sides by the reciprocal of $\frac{1}{7}$. And when we multiply a number by its reciprocal we get what number?"

Hearing "one," I now rewrite the equation: $\frac{7}{1} \cdot \frac{1}{7}y = 4 \cdot 7$.

"We changed the left side to $1y$, or y. We've isolated the variable and solved the equation. Can we now do the same thing to solve this equation?"

$$\frac{2}{5}x = 8$$

If a student connects the dots and says, "Multiply both sides by the reciprocal of the coefficient" I will celebrate the evening with champagne and caviar. But usually someone will say the familiar refrain that I just talked about: "Divide both sides by $\frac{2}{5}$," which is fine because I'll once again show that it's the same as multiplying both sides by the reciprocal: $8 \div \frac{2}{5} = 8 \times \frac{5}{2}$.

I then have them work examples to drive home the procedure.

Examples

1. $\frac{3}{7}x = 24$. Answer: $x = 24 \times \frac{7}{3}$, $x = 56$.
2. $-\frac{2}{9}m = 42$. Answer: $m = -\frac{9}{2} \cdot \frac{42}{1}$; $m = -189$.
3. $\frac{3}{5}t = 4$. *Answer*: $\frac{5}{3} \times \frac{4}{1} = \frac{20}{3}$ or 6 2/3.
4. $\frac{3}{7}x = \frac{4}{5}$. Answer: $x = \frac{4}{5} \cdot \frac{7}{3} = \frac{28}{15}$.

For these last two, I assure students that getting a fraction for an answer doesn't mean it's wrong.

Decimals as Coefficients

Equations such as $0.3x = 0.54$ and $\frac{x}{0.04} = 500$ are solved the same way as equations with integral coefficients. I start off by introducing a problem, such as the first one above, and asking how it would be solved.

Upon hearing "Divide both sides by 0.3" or something resembling it, I do so on the board, ending up with $0.54 \div 0.3$. Students know to move the decimal point of the divisor and dividend one place to the right, resulting in $\frac{5.4}{3}$, which is 1.8.

Homework

I assign problems with integer coefficients, as well as fractional and decimal coefficients. I also include problems solved by addition or subtraction.

For Curious Students

Problems such as $\frac{3}{7}x = \frac{4}{5}$ can be used to demonstrate why fractional division involves inverting the divisor and multiplying. I discuss this in a later unit in the chapter on algebra. For seventh grade, my explanation is less rigorous than the one I show in algebra.

The demonstration is an extension of the discussion for solving a problem with a fractional coefficient on the left-hand side and a whole number on the right. The demonstration is fairly straightforward. "We first solve this as a division problem. That is, we divide both sides by $\frac{3}{7}$, resulting in $x = \frac{4}{5} \div \frac{3}{7}$.

"Now, let's put that aside, and solve it by multiplying both sides by the reciprocal of $\frac{3}{7}$ which is $\frac{7}{3}$. We get $\frac{3}{7} \cdot \frac{7}{3} x = \frac{4}{5} \cdot \frac{7}{3}$ or $x = \frac{4}{5} \cdot \frac{7}{3}$."

"Since $\frac{4}{5} \div \frac{3}{7}$ and $\frac{4}{5} \times \frac{7}{3}$, both equal x they must be equal to each other."

"Therefore, $\frac{4}{5} \div \frac{3}{7} = \frac{4}{5} \times \frac{7}{3}$."

I may do this during the review for the test, telling students that I will give extra credit for students who can replicate the demonstration.

9. SOLVING WORD PROBLEMS USING EQUATIONS

PRIOR CONTENT

Students have learned the basics for two-step equations, as well as how to solve two-step equations with fractions, such as $\frac{1}{2}x + \frac{2}{3} = \frac{5}{6}$. This equation can be solved by expressing fractions with a common denominator on one side and then solving. For example, the above problem can be solved in this way: $\frac{1}{2}x = \frac{5}{6} - \frac{4}{6}$; $\frac{1}{2}x = \frac{1}{6}$; $x = \frac{1}{6} \cdot \frac{2}{1} = \frac{1}{3}$. Alternatively, it can be solved by eliminating the denominator by multiplying each term by the common denominator. The latter method proves confusing for some students; others take to it.

THIS LESSON

Word problems are admittedly hard for most students. It requires representing the problem in algebraic symbols. I start with problems that translate easily into symbols, and then progress to more complex problems. One difficulty with starting simple is that some of the problems may be solved without resorting to writing an equation. For example, the problem "What number added to 3 is 10?" I tell students that although the problem can be solved without an equation, they have to start somewhere. They need to be able to express the problem as $x + 3 = 10$. Eventually they will encounter problems that do not lend themselves to being solved quite so easily and an equation helps. This lesson also includes solving problems with two unknowns.

Warm-Ups

1. Solve. $\frac{2}{3}x + \frac{3}{5} = \frac{7}{10}$. *Answer*: $\frac{2}{3} = \frac{7}{10} - \frac{6}{10}$; $\left(\frac{2}{3}\right)x = \frac{1}{10}$; $x = \frac{3}{20}$.
2. $6 = -5x - 4$. *Answer*: $-5x = 10$; $x = -2$.
3. Translate into an algebraic equation and solve. Five added to 2 times some number is 25. *Answer*: $5 + 2x = 25$; $2x = 20$; $x = 10$.
4. $-6 - 2 = ?$ *Answer*: -8.
5. The average of two numbers is 23. What is the total of the two numbers? *Answer*: *If x is the total,then* $\frac{x}{2} = 23$; $x = 46$.

Problem 1 is a review of a two-step equation with fractions. Problem 2 may cause students difficulty because the variable term is negative. One prompt might be "What do you divide both sides by? What is the coefficient of *x*?" Problem 4 reminds students that the sum of two negatives is negative. Some students may still conflate this with the multiplication rule, and write the answer as positive 8. Problem 5 generally proves difficult for students. Possible prompt: "Can you let *x* be the total? What do we divide the total by to find the average?"

One Unknown

I start the lesson by giving them a problem. "I'm going to give you a problem that you will think you have the answer to and I'm willing to bet that it is wrong." This generally gets their attention.

"John and his sister have \$110 between them, and John has \$100 more than has sister. How much do each of them have?"

Usually the answer I get is \$100 and \$10, and I point out that it is wrong because 100 is 90 more than 10, not 100. Soon students resort to "guess and check" and come up with the correct answer: \$105 and \$5.

"Many of you solved this by guessing and checking, which is one way to solve problems, but it can be time-consuming. Today we'll learn how to solve word problems using algebra, and later how algebra is used to solve this problem. Now, we'll work on some simple word problems in your notebooks. The first step in a word problem is to identify what you're solving for, and identify that unknown quantity with a variable."

I give them a few examples to start with, working through the first two with them; for the rest, they work independently, and I require them to write the equation and solve it in their notebooks or on mini-whiteboards. Simply writing the answer will not suffice.

Examples

1. A certain number plus 8 equals 15. Find the number. *Answer*: *Let* x = *the number*; $x + 8 = 15$; $x = 7$.

I advise students to write down what the variable represents.

2. Two more than 3 times a number is 8. Find the number. *Answer: Let* x = *the number*; $2 + 3x = 8$; $3x = 6$, $x = 2$.
3. Eight less than 3 times a number is -23.

   Answer: $3x - 8 = -23$; $3x = -15$; $x = -5$.

Some students will write $8 - 3x$ rather than $3x - 8$.

4. Ten increased by $\frac{1}{6}$ of a number is 5.

   Answer: *Let* x = *the number*; $10 + \frac{1}{6}x = 5$; $\frac{1}{6}x = -5$; $x = -30$.

In summary, the following steps are followed:

Step 1. Read the problem to determine what value is unknown in the problem.

Step 2. Define a variable to represent the unknown quantity.

Step 3. Translate into an equation.

More than One Unknown

"Some problems have more than one unknown. For example, suppose a problem says, 'Five more than a number added to the number equals 60.' We have two unknowns here: a number and 5 times the number. How can I represent 'a number?'"

Response: x.

"How do I represent 5 times that number?"

Response: $5x$.

"So we have two numbers: x and $5x$. How do I represent that their sum equals 60?"

Hearing $5x + x = 60$, I ask students to solve it. "Can you combine terms?"

They do and get $6x = 60$. "Solve it." They do and tell me that x equals 10.

"That's correct, but the problem asked for two numbers. How do I find the other?" Someone will say to plug 10 into $5x$, which gives 50.

I do a few more problems as necessary until I see that students are comfortable with translating, and then I ask them to work on the examples.

EXAMPLES

1. Bob weighs 15 pounds more than Alan. Together, they weigh 99 pounds. How much does each boy weigh?" *Answer*: *Let* x = *Alan' s weight*; *Bob' s weight* $= x + 15$. $x + 15 + x = 99$; $2x + 15 = 99$, $2x = 84$; $x = 42$, $x + 15 = 57$.
2. Joe is 2 years older than Ed. The sum of their ages is 28. Find the age of each boy. *Answer*: *Let* x = *Ed's age*; *then Joe's age* $= 2 + x$; $x + 2 + x = 28$; $2x + 2 = 28$; $2x = 26$; $x = 13$, $x + 2 = 15$.

   Students will sometimes ask, "What if I had let x equal Joe's age?" In that case, Ed's age would be represented as $x - 2$; then, $x = 15$, and $x - 2 = 13$.
3. A certain number plus twice that number is –24. Find the two numbers.

   Answer: $x + 2x = -24$; $3x = -24$, $x = -8$.

Now, I ask students to solve the problem I gave them earlier, using an equation: "John has \$100 more than his sister. Together, they have \$110. How much does each person have?"

Prompts may include: "If x represents how much his sister has, how do I represent John's amount in terms of x?" I check notebooks and pick a student with the correct equation and solution to present it on the board. The equation is $x + 100 + x = 110$. Solving the equation, $x = 5$ and $100 + x = 105$.

Homework

I usually make up a worksheet of about 15 problems. I divide them up into three groups of five problems: one unknown, two unknowns, and mixed problems. The third group entails reading the problem through to determine whether it has one or two unknowns. I spend time going over the solutions with students the next day. Problems should ultimately be in the format of variables on one side of the equation only.

10. SOLVING INEQUALITIES WITH MULTIPLICATION AND DIVISION—NEGATIVE FACTORS

Following the lesson on word problems, we then cover solving inequalities with addition, subtraction as well as multiplication and division by positive numbers. Students are familiar with the basics of inequalities, having learned them in sixth grade. Nevertheless, phrases such as "no more than 50" and "at least 100" prove tricky.

This lesson introduces inequalities when both sides are multiplied or divided by a negative number, which reverses the sign of the inequality. In the lessons on inequalities, students continue solving word problems. For that matter, I include word problems all the way through the course, as well as equations and inequalities. Mastering the basic operations of algebra makes it easier for students when they take algebra. Students will see algebra as a continuation of what they started, rather than as a totally foreign subject.

WARM-UPS

1. The average weight of Steven and Allison is 75 pounds. What is their total weight? *Answer*: *Let* x = *total weight.* $\frac{x}{2} = 75$; $x = 150$.
2. Solve. $2x - 60 \leq 90$. *Answer*: $2x \leq 150$; $x \leq 75$.
3. Simplify. $-2(x - 5)$. *Answer*: $-2x + 10$.
4. Solve. $-2(x - 5) = 20$. *Answer*: $-2x + 10 = 20$; $-2x = 10$; $x = -5$.
5. Solve. $2x + 10 \geq 20$. *Answer*: $2x \geq 10$; $x \geq 5$.

Problem 1 is similar to a Warm-Up problem they had in an earlier lesson, which asked for the total of two numbers with a given average. Most students will now solve it by multiplying by 2 rather than writing it as an equation. Problem 3 helps set the stage for Problem 4, since the answer to Problem 3 is used in Problem 4.

Multiplying and Dividing by a Negative Number

After going over Problem 5 of the Warm-Ups, I tell students to change the problem slightly. "I want to change the problem to the following:"

$$-2(x - 5) \leq 20$$

"Let's distribute the –2, which you've already done in Problem 4 of the Warm-Ups."

They get $-2x + 10 \leq 20$.

"Now, isolate the variable term."

It is now $-2x \leq 10$. "Finish it and let me see what you get," I say, walking around and checking notebooks. What I see is $x \leq -5$.

"You've told me that the answer is $x \leq -5$. So, let's check this to see if it's true. Someone give me a number that is less than –5."

Hearing, for example, −6, I then plug in −6 for x in the original inequality.

$$-2(-6 - 5) \leq ?\ 20$$

"What happens inside the parentheses? What is −6 − 5?" Hearing –11, I then write:

$$-2(-11) \leq ?\ 20$$

"Is this true?"

Students see that $22 \leq 20$ is an untrue statement. "What went wrong?" I ask. "Let's take a closer look."

I write $-3 < 4$ and $7 < 9$ on the board and ask two students to help me graph the two pairs of points:

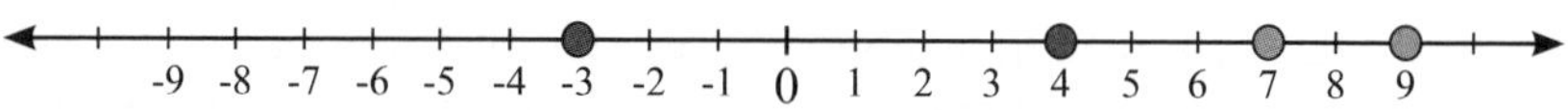

$$-3 < 4 \qquad 7 < 9$$

"For −3 < 4, what do these numbers become after I multiply by −1?"

Response: They are now 3 and −4.

"What inequality sign goes between the two numbers now?"

It becomes obvious that the sign is reversed as it is when 7 < 9 is multiplied by −1. I then have two more students plot the new points on a number line.

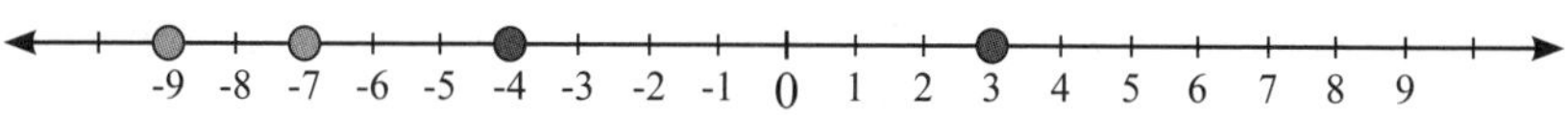

$$-3(-1)?4(-1):3>-4$$

$$7(-1)?9(-1):-7>-9$$

The graph makes it obvious what happens; the new numbers obtained by multiplying by −1 are a mirror image of the original points and are reversed.

Demonstrating this for division follows: For 6 > 2, dividing by −2 results in −3 < −1.

"Stated simply, multiplying or dividing both sides of an inequality by a negative number reverses the inequality sign." I ask students to write the rules in their notebooks:

If $a < b$ and $c < 0$ (c is negative), then $a \cdot c > b \cdot c$ *and* $\frac{a}{c} > \frac{b}{c}$.

If $a > b$ and $c < 0$ (c is negative), then $a \cdot c < b \cdot c$ *and* $\frac{a}{c} < \frac{b}{c}$.

Examples. Students write the answers in their notebooks.

1. Multiply both sides by −2: 5 < 8. *Answer* −10 > −16.
2. Multiply by −5: −3 > −4. *Answer*: 15 < 20.
3. Solve for x: $-3x > 15$. *Answer*: Dividing both sides by −3, we get $x < -5$.
4. Solve for y: $-\left(\frac{1}{2}\right)y \leq 5$.

Answer: Multiply by the reciprocal of $-\frac{1}{2}$; $y \geq 10$. Solve for x: $6x < -12$.

For the next step, many students will assume that the inequality sign reverses. I ask, "Are we dividing by a negative number?" No, we are not. Therefore, by the rule, the inequality sign remains the same.

Students catch on quickly, but I make sure to mix up problems of multiplication or division by a negative number as well as a positive number to ensure they make the distinction. I include some word problems as well:

A cab ride costs \$3 per mile. If a ride costs no more than \$21, what is the range of miles the ride can be? *Answer*: $3x \leq 21$; $x \leq 7$.

The answer indicates that the range can be up to 7 miles. That is, a cab ride must be 7 miles or less for the price to stay under \$21. It is important that students understand that inequalities represent a set of infinitely many answers that solve the problem. This solution set comprises a range of numbers.

Example. A number multiplied by –4 is greater than 12. Find all possible values of the number.

Answer: $-4x > 12$; $x > -3$.

All possible values then are all numbers greater than –3. Graphing this on a number line helps students understand that all numbers designated by the arrow to the right of –3 (but not including –3) satisfy the inequality.

Homework

Problems should include multiplying/dividing by negative numbers, as well as problems in which the inequality sign remains the same; that is, the right-hand side is negative, but the coefficient on the left-hand side is positive. I also include word problems.

11. RATIOS AND TAPE DIAGRAMS

PRIOR CONTENT

After the unit on algebraic expressions and equations, the next unit is ratios and proportions—a mainstay of seventh-grade math. Students have learned that a ratio is a comparison of two quantities by division. I present as a first example, the "Surf City" ratio, taken from the 1963 song of the same name (which none of my students has ever heard of): Two girls for every boy, or 2:1.

Ratios can be expressed in various ways. A paint recipe of 2 parts yellow to 3 parts blue can be written as 2:3, $\frac{2}{3}$, or 2 to 3. The paint recipe is a part-to-part ratio—these are ratios that compare the parts that make up the whole. Other examples include girls to boys in a classroom and the ratio of chocolate to non-chocolate candies in a box of candy.

A part-to-whole ratio compares a part of something to the entire quantity. For a classroom in which the number of girls to boys is in the Surf City ratio, or 2:1, the number of girls to total students in a classroom is 2:3; that's part to whole. This ratio holds no matter how many students are in the classroom and represents the fraction of students in the class who are girls. A typical problem includes "What is the fraction of girls in a class where the ratio of girls: boys is 6:7?" (Answer: 6: 13).

Lastly, students learn that a ratio of two quantities with different units is called a "rate." Examples include miles per hour, dollars per pound, and hours per day.

PRESENT LESSON

Having covered the basic structure of ratios, this lesson provides a problem-solving heuristic for ratios, which are sometimes referred to as "tape diagrams." Tape diagrams are a pictorial method employed in JUMP Math and in Singapore Math's books, both of which have served as my source for this technique. Tape diagrams serve as an entry point to algebraic solutions, which are addressed in the subsequent lesson. Students generally enjoy the pictorial method, which provides them a sense of accomplishment in solving problems that many students previously considered to lie well beyond their abilities.

WARM UPS

1. Tim's average score on the last four math tests is 92. If he gets a 97 on his next test, what is his average score? *Answer*: $92 \times 4 = 368$ *for the 4 tests*. $368 + 95 = 465$, *total score for* 5 *tests. Average score for* 5 *tests* $= \frac{465}{5} = 93$.
2. Harry bought 155 oranges for \$35. He found that 15 of them were rotten. He sold all the remaining oranges at 7 for \$2. How much money did he make?

   Answer: $155 - 15 = 140$ *good oranges*. $140 \div 7 = 20$ groups of 7 oranges.

   \$20 × 2 = \$40. \$40 − \$35 = \$5 *amount of money made.*
3. Mark makes 5 gallons of purple paint by mixing red and blue paint. He uses 2 gallons of red paint. What is the ratio of red to blue paint in his mixture? *Answer*: 2:3.
4. Write the sentence as an inequality and solve. "A number a divided by 2 is no more than 6." *Answer*: $\frac{a}{2} \leq 6$ or $\frac{1}{2}\, a \leq 6$; $a \leq 6 \cdot 2$; $a \leq 12$.
5. At the Cypress School, the total number of students is 300. Two hundred of them are girls. What is the ratio of girls: boys in simplest terms? *Answer*: 2:1.

For the first problem, students use what they know about finding the total amount prior to finding the given average. In this problem, they then have to add the fifth test score to obtain the total for five tests, and then divide by 5 to get the new average. Problem 2 is a multi-step problem, which is taken from Singapore Math's *Primary Math* series. My experience in giving this problem is that seventh-graders enjoy solving it, and they look at it as a puzzle. Problems 3–5 review previous material.

Using Tape Diagrams (Bar Models)

I start this lesson off by asking a question I know students know the answer to: "What is the Surf City ratio?" I usually hear in unison: "two boys for every girl."

I give the students a problem. If the total number of students in a school is 600 and we have a 2:1 (Surf City) ratio of girls to boys, how many girls and boys attend the school?

I draw a diagram representing the 2:1 ratio on the board:

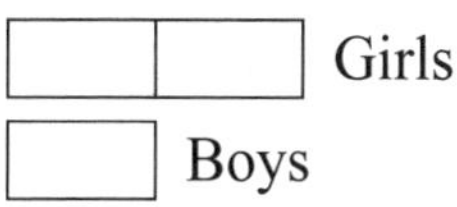

"Each box can represent any amount of students as long as all the boxes represent the same number. So, if a box equals 50, how many girls are there?" (I write 50 in each of the two boxes marked girls.) I'm told fairly quickly that there are 100 girls.

"And how many boys?"

Response: 50.

"And what is the total number of students?" There is a pause usually, and eventually I hear 150.

"Since the number in each box must be equal—and there are three boxes—how do we find the number in each box if the total for all three boxes is 600 as the problem states?"

Response: 200.

I will ask how they calculated it and am satisfied if the answer resembles "Divide the total number by the number of boxes to get the number in each box."

Examples: I provide examples for students to work on in their notebooks. Problems are projected on the wall.

1. I have red and green marbles in the ratio of 5:2, and a total of 28. How many of each do I have?

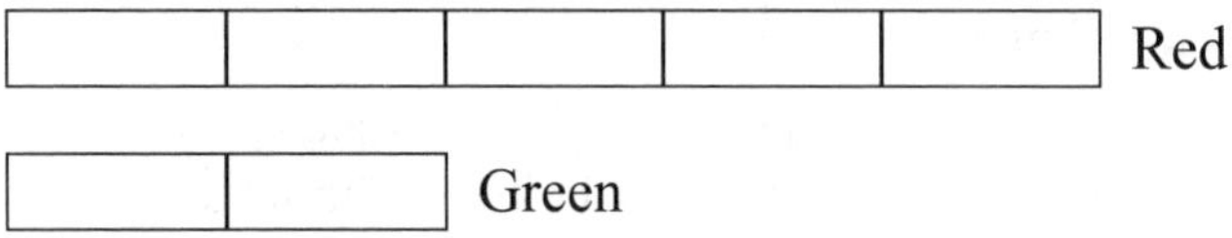

"What is the total?" I ask. 28.

"How many total boxes?" 7

"How many in each box?" A pause, and then I hear, "4"

I write 4 in each box and have the students tell me how many marbles are in each box:

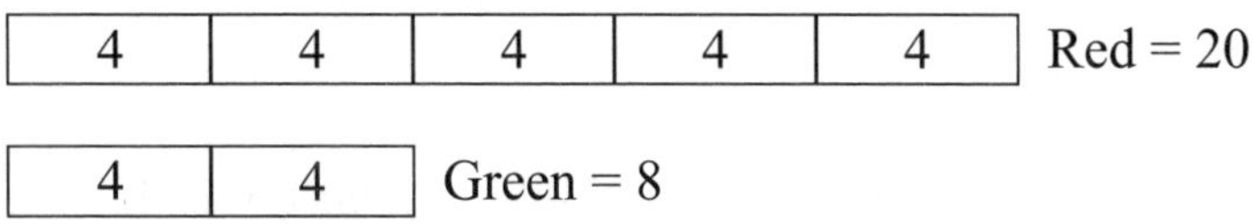

2. Use the same ratio as the first example, but the total is 35 marbles. How many are there of each color? *Answer: 5 in each box—25, with 10 red and green, respectively.*

"Is there a more efficient way of finding the total number of boxes than by counting them?" Usually someone will say that the sum of the two numbers in the ratio gives us the total.

3. Use the same ratio as before: 5:2. This time, the total is not given—only that there are 50 red marbles. How many green marbles are there?

"Any ideas on how to solve this one?" If I need to give a hint, I will say, "How many boxes represent the red marbles?" There are 5 boxes.

"Can you tell me then how many are in each box if there are 50 red marbles?" They see right away to divide 50 by 5: There are 10 marbles per box.

"How many green marbles are there then?"

Each box contains 10; thus, in the two boxes representing the green marbles, there are 10×2, or 20.

4. The ratio of girls: boys is 3:5. There are 30 boys. How many girls are there? I want them to draw the boxes and do the problem without hints from me.

   Answer: 30 boys, 5 boxes = 6 per box. Girls, 3 boxes: $6 \times 3 = 18$.

More than Two Ratios

"We can compare more than two things using ratios. Let's say we have blue, green, and white beads in the ratio of 5:2:3." I draw the boxes for the beads:

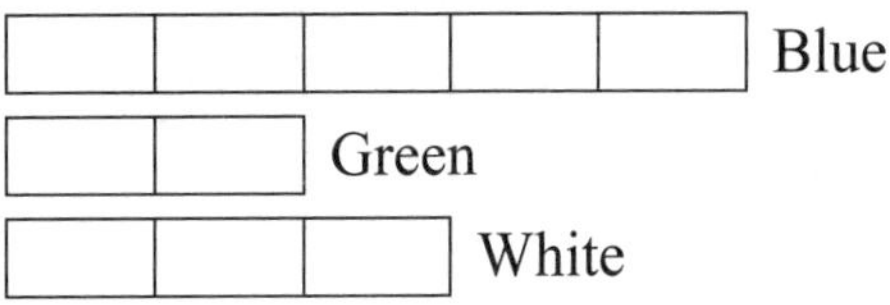

"If the total equals 180 beads, how many of each are there? We solve it the same way as when there were just two quantities; who can do this?" Generally, a prompt such as "How many boxes are there?" gets them going. Since there are 10 boxes, $\frac{180}{10}$ equals 18 beads per box.

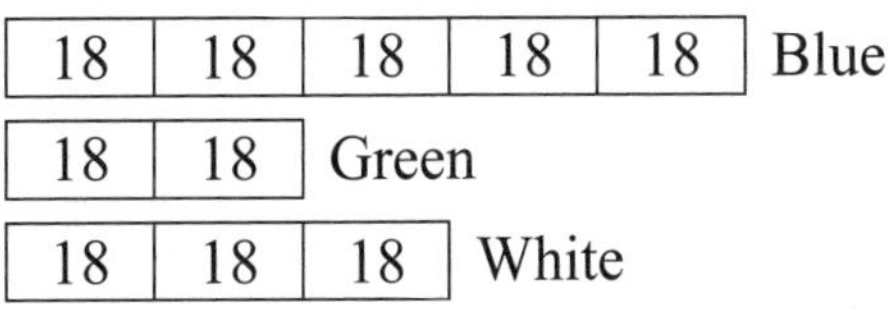

Blue: 90, Green: 36, White: 54

Examples

1. There are 3 apples and 2 oranges for each plum in a fruit stand. There are 420 fruits altogether. How many of each fruit are there? *Answer: Total of 6 boxes; 420/6 = 70, so there are 3 x 70 = 210 apples, 70 x 2 = 140 oranges, and 70 plums.*

The part that confuses students is how many boxes represent plums. Since the problem states the numbers of apples and oranges in terms of each plum, then one box represents plums.

2. To make purple paint, the colors red, blue, and white are mixed as follows: Two times as much blue as white. Three times as much red as white. If we want 30 gallons, how much of each of the colors do we mix?

Since the problem is stated differently than the others, I give students a prompt: "How do we draw the boxes for blue and white if blue is twice the amount of white?"

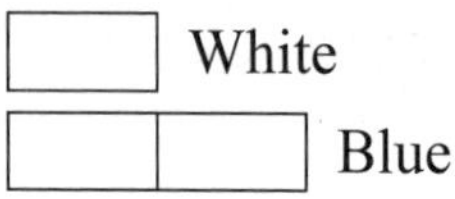

Following the same pattern, there are three red boxes with respect to white:

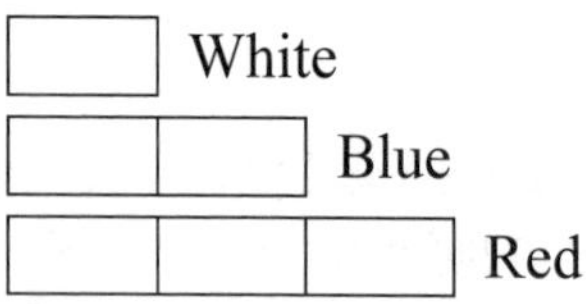

Now, the problem is easy. *Answer:* $\frac{30}{6}$ *= 5 per box. White: 5, Blue: 10, Red: 15.*

More of One Thing than Another

The last structure I cover are problems that indicate how much more of one thing we have than another. "Suppose we have the red and green marbles in a ratio of 5:2 again, but this time I tell you that there are 12 more red marbles than green marbles. I want to know how many of each color I have."

Before students have a chance to tune out because they think it's too hard, I immediately draw the diagram on the board. "We know the ratio is 5:2, so let's draw the boxes first and go from there."

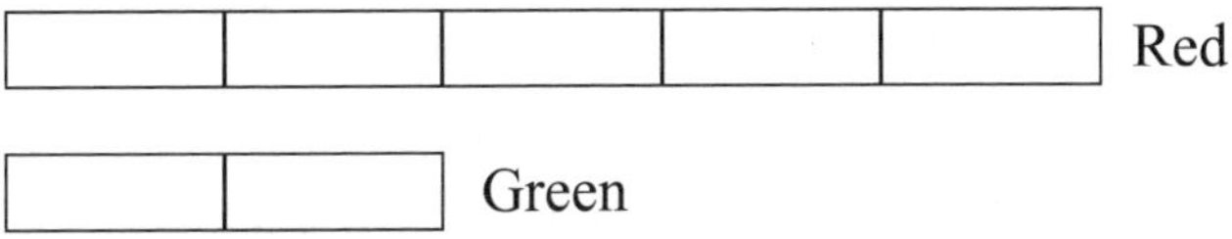

"The top row of boxes represents red marbles. How many more red boxes do we have than green boxes?"

Hearing "two," I ask, "Those two boxes represent how many marbles?"

Some students see right away that two boxes more of red than green represents 12 marbles. For those who don't see it, I refer them to the problem and the wording "12 more red marbles than green marbles." I'll hear the "Ohhh" of insight from those who understand. To make sure all understand, I then say, "And there are two more red boxes than green and those two represent how much?"

Response: 12.

There are then $\frac{12}{2}$, or 6, marbles in each of the boxes, giving us an answer of 30 red and 18 green.

"OK, same problem, different number: There are 40 more red marbles than green marbles. How many of each do I have? Figure it out in your notebooks."

I walk around to check on their work. Most students get it; I offer prompts for those having difficulty. $\frac{40}{2} = 20$ per box, giving us 100 red and 60 green.

Examples

1. The ratio of roses to tulips is 2:3. There are 10 more tulips than roses. How many of each are there? *Answer: Since the tulips are one box more than roses, each box = 10, and there are 20 roses and 30 tulips.*
2. At Tidewater Tech, the ratio of full-time teachers to part-time teachers is 5:3. There are 18 more full-time teachers than part-time. How many of each are there? *Answer: There are two more full-time teacher boxes than part-time, so each box* $= \frac{18}{2} = 9$. *There are* $9 \times 5 = 45$ *full time teachers and* $9{\times}3{=}27$ *part time teachers.*

Homework

I use problems that are in JUMP Math's *Assessment and Practice* books, a page of which I have reproduced below (see Figure 11-1). I work with the students on these problems. They find them fun and fairly easy, and they generally finish them in the remaining class time.

7. Find the number of cups of blue (b) and yellow (y) paint needed to make green paint.

a) blue paint : yellow paint = 4 : 5
45 cups altogether
b:
y:
blue: ______
yellow: ______

b) blue paint : yellow paint = 5 : 3
8 more cups of blue paint than yellow
b:
y:
blue: ______
yellow: ______

c) blue paint : yellow paint = 3 : 7
12 more cups of yellow paint than blue
b:
y:
blue: ______
yellow: ______

d) blue paint : yellow paint = 7 : 5
36 cups altogether
b:
y:
blue: ______
yellow: ______

e) blue paint : yellow paint = 4 : 7
33 more cups of yellow than blue
b:
y:
blue: ______
yellow: ______

f) blue paint : yellow paint = 5 : 2
42 cups of paint altogether
b:
y:
blue: ______
yellow: ______

8. Draw a model to answer the question.

a) There are 35 students in a class. The ratio of girls to boys is 3 : 2. How many girls and how many boys are in the class?

b) There are 44 marbles in a jar. The marbles are blue and red. The ratio of blue marbles to red marbles is 5 : 6. How many red marbles and how many blue marbles are in the jar?

c) Clara collects American and Canadian stamps. She has 6 more American stamps than Canadian stamps. The ratio of Canadian to American stamps is 3 : 5. How many of each kind of stamp does she have?

9. Peter is 6 times as old as Ella.

a) What is the ratio of Peter's age to Ella's age?

b) Peter is 15 years older than Ella. How old are Peter and Ella?

Figure 11-1: Problems from JUMP Math; Assessment and Practice 7.1, p. 23. Reprinted with permission.

12. SOLVING RATIO PROBLEMS ALGEBRAICALLY

The tape diagrams from the previous lesson now will be used as a gateway to solving the same type of problems using algebraic equations. Some students may find solving the problems easier using algebra. Others will be more dependent on the tape diagrams and will use them to derive the algebraic equations. The approach may be more applicable for accelerated seventh-grade math and Math 8 classes, but I teach it in regular Math 7 classes as well. In all such classes I allow use of algebra for ratio problems as an option and award extra credit on quizzes and tests for those who use it (and get the correct answer!). I've had students who, when they take algebra and have the same type of problems, remark that now they seem so much easier than they did in seventh grade.

In this lesson, I push for all students to at least try using algebra. For the homework, however, I allow students to use tape diagrams if they are unable to use algebra to solve a problem. Most students will make an effort to use algebra.

Warm Ups

1. Tim's average score on the last four math tests is 95. If on his fifth test his average is 96, what is the total of his five scores? What was his score on the fifth test? *Answer: Find the total for his current average: 4 x 95 = 380. For the score on his fifth test, find the total for the average of 96 with five tests: 5 x 96 = 480. 480-380 = 100, which would have to be the score on the fifth test.*

2. Peter saves \$52. Susan saves \$20 more than Peter. Find the ratio of Peter's savings to Susan's savings in the simplest terms. *Answer*: 52:72 =13:18 *or* $\frac{13}{18}$.
3. Express the ratio of 40 minutes to 2 hours in the simplest terms. *Answer*: 40:120 = 1:3.
4. The ratio of girls to total number of students in class is 5:12. There are 14 more boys than girls. Find out how many boys and girls there are. *Answer: Drawing the tape diagrams, there will be 7 more boxes representing students, which would represent the number of boys. Since there are 14 boys, 14 ÷ 7 = 2, and 2 goes in each box. Number of girls = 5 x 2 = 10; number of boys = 14 + 10 = 24.*
5. The ratio of the number of apples to the number of oranges is 7:4. There are 60 oranges. How many apples are there? *Answer: Since 4 boxes represents number of oranges and there are 60, then each box in the tape diagram equals 15. No. of oranges = 7 x 15 = 105.*

Problem 1 may give the students some difficulty. A helpful prompt is to ask what the totals of the scores represent—that is, what does the total of the five scores have that the total of the four scores does not? For Problem 4, the tape diagram will show 7 more boxes representing students. Ask what the additional 7 boxes represents—more girls or boys? Problem 5 prompt: "Which boxes would represent oranges? How many boxes are there?"

THE ALGEBRAIC APPROACH

I start out by giving a problem and having students draw the tape diagrams. The problem is projected on the wall:

There are 36 students at a dance, and the ratio of girls to boys is 4:5. How many of each are there?

I draw the tape diagrams in the usual manner.

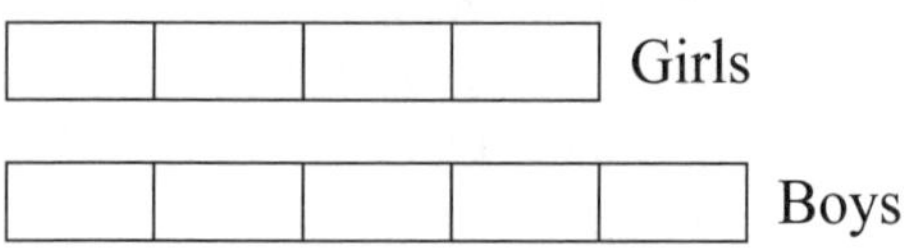

Students will usually go ahead and solve the problem.

"I can see that most of you have solved this. Someone tell me what you did."

Students tell me they put 4 in all the boxes and calculated the number of girls and boys. For girls, they multiplied 4 × 4. For boys, they multiplied 4 × 5.

"Correct; now I want to do it a little differently. Let's pretend that we don't know what goes in the boxes."

I have to say that in all the time I've been doing this, no one has ever said, "But we *do* know!" This isn't to say that it won't happen.

"When we don't know what a number is in a problem, what do we use?"

Students, as always, say "*x*." "I'm going to write x in all the boxes." I do this on the diagram I have on the board.

x	x	x	x	Girls

x	x	x	x	x	Boys

"For the number of girls, instead of multiplying a number by 4 boxes, what do we multiply by 4, and how do we write it?" Someone will come up with $4x$. (If they don't, I do my best imitation of Jaime Escalante in the movie "Stand and Deliver" and looking upward say, "Forgive them, they know not what they do." I did this in a Catholic school once. There was no response.

"We do the same thing for number of boys; how do I represent them algebraically?" I usually hear $5x$. I write on the board: $4x$ = *number of girls*; $5x$ = *number of boys.*

"The problem states that the total number of boys and girls equals what?" Hearing 36, I continue. "If the number of girls is $4x$, the number of boys is $5x$, and the total is 36, how do I write this as an equation?"

In case of no response, I write on the board: *Number of girls + number of boys* = 36.

More than one student will come up with the answer: $4x + 5x = 36$.

"Solve it," I say, and they do. Combining terms and solving: $9x = 36, x = 4$.

"We've found the number that we normally put in the boxes, but we did it with algebra. But we haven't solved the problem."

I point to the statement on the board defining what $4x$ and $5x$ represent. "In this problem, x is not the answer; it is the multiplier. Somebody tell me the number of girls."

Most see where I'm going with this, and I'm told it is 16. Similarly, the number of boys is $5x$, which equals 20.

Examples. We now work with examples, which I have them write in their notebooks as I go around to offer help and guidance as needed. I do the first one with the class.

1. The ratio of roses to tulips is 2:3. There are 10 more tulips than roses. How many of each are there?

I want to know how to represent the number of roses and the number tulips in terms of x.

"To help you do this one, draw a tape diagram and fill in the boxes with x." This helps them to define $2x$ and $3x$ as the number of tulips and roses, respectively.

"This is a 'More of one thing than another" type of problem. How do we solve this?"

After I hear a correct answer, I tell them to write this down:

Number of tulips – number of roses = 10

"You wrote down the number of roses and tulips in terms of x, so let's substitute them in the sentence to make it an equation that we can solve."

$$3x - 2x = 10; x = 10$$

After solving for x, I ask: "Have we solved the problem?"

Some students will say "yes" and some "no."

I make clear that we want to know the number of tulips and roses and that x tells us only what the multiplier is.

"$3x$ is the number of tulips, so what does that equal? What is the number of roses?" We soon have the answer: 30 tulips and 20 roses.

2. The ratio of girls to boys at a football game is 3:5. There are 30 boys. How many girls are there?

For this, I provide initial guidance. "Which number, $3x$ or $5x$, is going to help us here? What is 30 in the problem? Boys or girls?"

I'm trying to lead them to $5x = 30$. Solving it yields $x = 6$. The number of girls is given by $3x$, which is 18.

3. An 80-meter cable is cut into two pieces. The lengths of the pieces have a ratio of 11:5. How long is the shorter piece?

Answer: $11x + 5x = 80$; $16x = 80$; $x = 5$.

Shorter piece: $5x = 25$ meters.

4. There are 35 students at a game. The ratio of boys to girls is 4 to 3. How many boys are there?

Answer: $4x + 3x = 35$; $7x = 35$; $x = 5$. Boys: $4x = 20$. Girls: $3x = 15$.

"You may be wondering why we would bother with algebraic equations when we can draw tape diagrams."

There is usually some agreement here, so I give them the following problem and ask which method they think would be easier:

"The ratio of gallons of oil to gallons of water in a container is 33:78 and totals 333 gallons. How many gallons of each are there?"

I work through it with the class. 33x represents gallons of oil; 78x represents gallons of water. The equation is $33x + 78x = 333$; $111x = 333$, and $x = 3$. Therefore, there are 33×3 or 99 gallons of oil, and 78×3 or 234 gallons of water in the container.

There is general agreement that the algebraic method is easier.

Homework

As mentioned earlier, some students will rely on the tape diagrams to translate the problem into equations. For that reason, on the homework worksheet I include at least one problem with larger numbers, like the one above.

13. SOLVING PROPORTIONAL RELATIONSHIPS

Students have learned about unit rates in a previous lesson (not included in this book). Usually unit rates are used for comparisons; such as which is the better deal, 4 pounds of oranges for \$6, or 5 pounds for \$8?

Problems that are solved using unit rates can also be solved using more general proportion techniques. For example, students learn to solve this problem using unit rates:

"A bakery can make 195 donuts in 3 hours. At this rate, how many doughnuts can the bakery make in 8 hours?"

To solve this, students first find the unit rate: $\frac{195}{3}$ =65 donuts per hour. They then multiply the unit rate by 8. In today's lesson, students learn to solve it by the more general approach of equivalent ratios which is the topic of this lesson: $\frac{195}{3} = \frac{x}{8}$. In fact, the above problem when solved using the general approach encompasses the unit rate approach, if the left-hand side is simplified to $\frac{65}{1}$. Unit rates work well when numbers divide evenly as in this example; that isn't always the case, however.

My emphasis in teaching ratios and proportions for seventh grade has been focused on the fundamentals of proportions. Common Core standards require students to go beyond the fundamentals and connect unit rates to 1) the constant of proportionality, 2) the slope of graphs of proportional relationships, and 3) the equation form of proportional relationships (i.e., $y = kx$).

I cover these topics because they are in the Common Core standards. I do not dwell excessively on them, however. At that point, the concepts of constant of proportionality and direct variation make more sense in an algebra course after students have mastered the foundational aspects of proportions. The equation $y = kx$ follows as a consequence, in which y is the dependent variable, x the independent variable, and k the constant of proportionality.

In the next chapter, I summarize what I teach regarding slopes, graphs of proportional relationships, and the equation form of direct proportions.

WARM-UPS

1. The lengths of 3 rods are in the ratio 1:3:4. If the total length is 96 cm, find the length of the longest rod. (Solve using algebra; tape diagrams may be used.) *Answer: Longest rod is* $4x.x + 3x + 4x = 96$; $8x = 96$; $x = 12$, so $4x = 48$ *cm.*
2. The average cost of 3 books is \$4.50. The average cost of two of the books is \$3.90. Find the cost of the third book. *Answer: Find the total cost for the two books: 3.90 × 2 = 7.80. Find the total cost for the 3 books: 4.50 × 3 = 13.50. The cost of the third book is therefore \$13.50 - \$7.80 = \$5.70.*
3. If 4 gallons of gasoline cost \$14.00, what is the cost of 10 gallons? (Use unit rate to solve). *Answer: Unit rate:*14/4= \$3.50/*gal.* $\frac{\$3.50}{gal} \times 10\ gal = \$35.00$.
4. Simplify. $-5x + 3y - 2x - 4y$. Answer: $-7x - y$.
5. Richard rowed a canoe $3\frac{1}{2}$ miles in $\frac{1}{2}$ hour. What is his speed in miles per hour? *Answer*: $3\frac{1}{2} \div \frac{1}{2} = \frac{7}{2} \times \frac{2}{1} = 7$ *hours.*

For Problem 1, students may use tape diagrams to write the algebraic equations or may use them with numbers. Problem 2 might prove difficult. Prompts include "Can we find total cost for the 2 books? What about 3 books?" Problem 3 is a unit rate problem. Students may express the unit rate in gallons/dollar rather than dollars/gallon. Problem 5 is also a unit rate problem, except that the numbers are fractions. A hint might be "What if he went 6 miles in 2 hours? How could we express that as a unit rate of mi/hr?" Students will generally find whole numbers easier to work with—it is then a matter of applying the same operation (division) to the fractions.

What a Proportion Is

I ask the class what the Surf City ratio is. If they have learned nothing else in this particular unit, they seem to grab on to this particular ratio and will call out "two girls for every boy," after which I write: $\frac{2}{1}$ on the board.

"Do 6 girls and 3 boys form the Surf City ratio?" To the chorus of "yes" that erupts, I then ask, "How do you know?" Students have little difficulty showing that $\frac{6}{3}$ is the same as $\frac{2}{1}$.

"How about ratios of girls to boys of $\frac{10}{5}$, $\frac{8}{4}$, $\frac{12}{6}$?" Students tell me they are all the same.

"Ratios that simplify to the same ratio like these are called 'equivalent ratios,' which bring us to the definition of 'proportion.'" I ask students to write the definition in their notebooks:

*A **proportion** is an equation stating that two ratios or rates are equivalent.*

For example: $\frac{6}{4} = \frac{12}{8}$; $\frac{7}{5} = \frac{14}{10}$.

Taking a look at the first equation, I try something. "What is the common denominator of 8 and 4? Not the lowest—a common denominator."

Hearing 32, I say, "Let's multiply both sides of the equation by 32 and see what happens."

$$\frac{6}{4} = \frac{12}{8}$$

$$\frac{32}{1} \cdot \frac{6}{4} = \frac{12}{8} \cdot \frac{32}{1}$$

I write $\frac{32}{1}$ rather than 32 because students will stubbornly refuse to see that a whole number can be multiplied by a fraction without writing the whole number as a fraction with denominator of one.

Next, we cross cancel, getting the following:

$$8 \cdot 6 = 12 \cdot 4$$

This yields an important result. "Notice that the 6 in the numerator on the left-hand side of the original proportion is multiplied by the 8 in the denominator on the right-hand side. What do you notice about the right-hand side?" At least one student will notice that the 12 in the numerator on the right side is multiplied by the denominator of 4 on the left side.

"This is called 'cross multiplication,' which is accomplished by multiplying diagonally as I just did. If both sides are equal, we'll get the same number on each side. Which we did: 48. That tells us that the two ratios are equivalent and therefore proportional. Try it with these and see if you get the same number on both sides."

$$\frac{3}{5} =? \frac{4}{9}$$

Students do this and obtain 27 and 20. "They are not equal, so we can say that these two ratios are not proportional. All this is an informal way to show this formal statement, which you should write in your notebooks.

$$If\ \frac{a}{b} = \frac{c}{d}\ then\ a \bullet d = b \bullet c$$

Using Cross Multiplication to Solve Problems

"The process we just described is actually very useful; we can use it to solve proportion problems."

"For example, green paint is made with 3 parts yellow and 2 parts blue. If we use 12 quarts of yellow how many quarts of blue must we use to make green paint?"

I then write the proportion with the two ratios of yellow to blue that we're interested in: $\frac{3}{2} = \frac{12}{x}$. "To solve for x we cross multiply like we did before." I draw arrows to show what it is we're multiplying.

"We multiply 3 by x, and 12 by 2. How do I write 3 times x?"

I wait until I hear 3x and then write it down. 12 times 2 they know.

"The result is the equation $3x = 24$. Solve it."

I wait for them to solve it and write the answer (8 in this case) on their mini-whiteboards.

"It's easy to check. We just cross multiply: $\frac{3}{2} = \frac{12}{8}$ gives us 24 = 24. Correct."

The above example is a part-to-part problem—yellow to blue. But sometimes problems are part to whole. I give the class this problem:

"To make green paint, the ratio of yellow to blue paint is 3:2. How much yellow paint is needed to make 20 gallons of green paint?"

I ask: "Is this problem part to part or part to whole?"

In case there is no response, I offer a prompt: "Are we asking how much blue paint to add like the first problem? Or are we looking at the total mixture that would be green paint?" It is then fairly obvious that it is part to whole.

"Since the ratio of yellow to blue is 3:2, what is the ratio of yellow to green?"

In the case of blank stares, I offer another prompt: "If we were talking about 3 gallons of yellow and 2 gallons of blue, what would be the total amount of green?"

It is now obvious to most that the ratio is 3: (2 + 3) or 3:5. I'll ask the students how the equation should be set up and ask them to write it on their mini-whiteboards. I'm looking to see: $\frac{3}{5} = \frac{x}{20}$. Initially, it helps if they give themselves a guide; I write on the board:

$$\frac{yellow}{green} = \frac{3}{5} = \frac{x}{20}$$

This can help keep numbers in the right order. Solving it: 5x = 60, and x = 12 gallons of yellow paint.

Now, I offer a problem with a rate. They have learned previously that rates are a ratio of two quantities that are different units, such as miles per gallon or dollars per pound.

"The cost of 5 oranges is \$2. How much will 15 oranges cost?" I ask students to write on their mini-whiteboards and am looking to see:

$$\frac{cost}{no.\ of\ oranges} = \frac{2}{5} = \frac{x}{15}\ OR\ \frac{oranges}{cost} = \frac{5}{2} = \frac{15}{x};\ 5x = 30;\ x = \$6$$

I add that for unit costs we generally write the ratios the first way: cost per item. But I show that in writing a proportion equation, we can write it the other way as well. As long as both sides of the equation are consistent, either way will produce the same answer.

EXAMPLES

1. "The ratio of Jim's weight to Eva's weight is 5:4. If their total weight is 90 lb, how much does Eva weigh?"

Some students may see that the problem could be solved as was done for ratios in a previous lesson: $5x + 4x = 90$; $9x = 90$, $x = 10$, and $4x = 40$.

2. "The ratio of boys to girls in a school choir is 4:3. If there are 16 boys, how many girls are there?"

   Answer: *Part to part*: $\frac{girls}{boys} = \frac{4}{3} = \frac{16}{x}$. $4x = 48$; $x = 12$.

3. "John reads 8 books in 12 weeks. If he keeps reading books at this rate, how many books will he have read in 15 weeks?"

   Answer: *This is a rate problem*: $\frac{books}{weeks} = \frac{8}{12} = \frac{x}{15}$.

   Note that $\frac{8}{12} = \frac{2}{3}$ *so* $\frac{2}{3} = \frac{x}{15}$, $3x = 30$, $x = 10$ *books.*

4. A car travels 300 miles in 5 hours. How far does it travel in 2 hours?

   Answer: *Rate problem*: $\frac{distance}{time} = \frac{300}{5} = \frac{x}{2}$. $\frac{300}{5}$ *reduces to* $\frac{60}{1}$.

   $\frac{60}{1} = \frac{x}{2}$, $x = 120$.

HOMEWORK

Textbooks provide a variety of problem. Some problems can be solved using the proportion method or using other methods, as seen in Example 1 above.

14. OTHER ASPECTS OF PROPORTIONS

Rather than provide a description of a lesson, I present here a summary of key points that I teach regarding 1) the constant of proportionality, 2) the equation form of proportional relationships (i.e., $y = kx$), and 3) graphs of proportional relationships. For accelerated seventh-grade math, I also cover slopes of lines, and how the slope relates to proportional relationships; specifically, that the slope of a proportional equation equals the unit rate. The coverage of these topics is primarily procedural, given that students will cover these topics in greater depth in algebra, building on their foundation in working with proportions.

Constant of Proportionality

The quotient of equivalent ratios (e.g., $\frac{3}{2}, \frac{6}{4}, \frac{9}{6}, \frac{12}{8}$) in lowest terms is called the "constant of proportionality."

If the ratio of cost($)/pound of oranges at various stores were $\frac{6}{4}, \frac{10}{5}, \frac{15}{6}$, it is easily seen that the ratios are not equivalent: $\frac{3}{2}$, 2, and $\frac{5}{2}$ are not equal. When two or more ratios are not constant, the quantities are *non-proportional.*

When working with rates such as dollars per pound, as in the above example, students learn that the constant of proportionality is equal to the unit rate. If it costs $14 for 4 gallons of gasoline, the unit rate is $\frac{14}{4}$, which in this case is $3.50 per gallon.

Writing Formulas or Equations Using Unit Rates

Proportional relationships can be described by equations of the form $y = kx$, where k is the constant of proportionality or, when rates are involved, the unit rate. The y variable is dependent on what x equals.

A proportion of unit rate can be expressed as an equation in this form. Look at the example of gasoline at \$3.50/gal. We don't need to always use y and x as variables. We can let c represent the cost of gasoline, and g the number of gallons. To put this in equation form, I teach students to use the following procedure:

Step 1: Express the unit rate as a proportion: $\frac{c}{g} = \frac{3.5}{1}$.

Step 2: Cross multiply: $c = 3.5g$.

The resulting equation is now in the form of $y = kx$. The constant of proportionality or unit rate for the equation is \$3.50, written as 3.5. The equation can be used to calculate the cost of gasoline for any number of gallons, represented by the variable g. I ask students to use it to calculate the cost of 20 gallons. "It's a plug in," I'll remind them. We then have $c = 20 \cdot 3.5 = \$70$.

GRAPHING PROPORTIONAL RELATIONSHIPS

Students were introduced to graphing in sixth grade, and perhaps even in fifth grade. The conventions of x and y axes and displaying ordered pairs as (x,y) are reviewed as part of the lesson. After graphing proportional and non-proportional relationships, the two are compared, and the rule for the graphs of proportional relationships is given:

If the graph of two related quantities is a straight line through the origin, then the two quantities are proportional. If it does not pass through the origin, the two quantities are not proportional.

Textbooks usually include a section on analyzing proportional relationships, in which the focus is on interpreting what the relationship is telling us. Their method is often along the lines of "The point $(1,y)$ tells what the constant of proportionality or unit rate is." For graphs that do not pass neatly through the x-coordinate of 1, the constant of proportionality or unit rate is found by dividing any y-value by its corresponding x-value—providing the relationship is indeed proportional. If all the $\frac{y}{x}$ ratios in a table are constant, the relationship is proportional.

15. EQUATIONS WITH VARIABLES ON BOTH SIDES

After the unit on ratios and proportions, we move into a unit on percents. Before getting into the unit, I revisit solving equations, focusing on more complex situations.

Normally this topic is addressed in accelerated seventh-grade math, but I have included it in regular classes. Both classes have difficulty at first. Not everyone in the regular math class catches on, but in such classes I don't put a large amount of weight on the topic in terms of tests as I do in the accelerated class.

One difficulty students have with these type of equations is organizing their steps in a way that they can read them and know where they are in the problem. In this lesson, I provide an orderly way that doesn't take up a lot of space and allows students to retrace their steps as necessary.

Warm-Ups

1. Solve. $3x + 2 = 8$. *Answer*: $3x = 6$; $x = 2$.
2. Solve. $\frac{2}{3}m = 10$. *Answer*: $m = 10 \times \frac{3}{2} = 15$.
3. Solve. $\frac{2}{3}t - 5 = 15$. *Answer*: $\frac{2}{3}t = 20$; $t = 20 \times \frac{3}{2} = 30$.
4. Write an equation and solve. Four less than three times a number is 8. *Answer*: $3x - 4 = 8$; $3x = 12$; $x = 4$.
5. Solve. $4x = 3x + 9$. *Answer*: $x = 9$.

All of the problems are two-step equations. Problem 5 is one in which there are variables on both sides. There will be questions on this, and a typical hint might be "How can I isolate the 9 on the right-hand side? How would I eliminate 3x from that side?" If they are still stuck I ask, "How did you get rid of the 2 on the left-hand side in Problem 1? Can you do the same thing with 3x?" If they ask, "Can you do that?" I respond, "Why not?"

OVERALL APPROACH

Problem 5 of the Warm-Ups serves as the conduit to the day's lesson. In going over the problem, I explain that they had difficulty with this only because it was something they hadn't seen before.

"The name of the game with problems like these is to get the variables on one side of the equation, and numbers on the other. These problems are similar to two-step equations except now you add or subtract variables as well as numbers.

I write a problem on the board and ask them what I should do: $9x - 24 = 3x$.

"We want to get the variables on one side. What should I do?"

This is a similar enough problem to the Warm-Ups, so students should see they can subtract $3x$ from both sides:

$$\begin{array}{r} 9x - 24 = 3x \\ \underline{-3x = -3x} \\ 6x - 24 = 0 \end{array}$$

I show how they can write the steps in a different way that takes up less space.

Students can write it like this: $9x - 24 - 3x = 3x - 3x$.

The $3x - 3x$ on the right-hand side will still seem strange at first, but with more problems like this, they get used to it. It also eliminates the habit some students have of writing the term to be added or subtracted under the wrong term in the equation—which causes them confusion.

The next step is then to add 24 to both sides using this same method of writing down the step:

$$6x - 24 + 24 = 0 + 24$$

The resulting equation, $6x = 24$, is easily solved.

I use a few more examples of this type:

$$12 - y = 5y$$

"What should we do first?"

Ideally, someone will suggest adding y to each side:

$$12 - y + \mathrm{y} = 5y + y;\ \textit{result}\text{: } 12 = 6y$$

After they solve it, I ask, "Let's see what happens if we subtract 5y from each side first rather than adding y to each side." I do this on the board.

$$12 - y - 5y = 5y - 5y.$$

"Help me combine terms," I say. In the interest of keeping the flow going, I pick students who I am certain will tell me correctly what to do:

$$12 - 6y = 5y - 5y$$

$$12 - 6y = 0$$

"What do I do now? Should I subtract 12 from each side, or add $6y$ to each side?"

I'll have one student subtract 12 from each side, and another add $6y$ to both sides. We end up with two equations:

$$-6y = -12 \textit{ and } 12 = 6y$$

For the first equation, we divide each side by -6. For the second, we divide each side by positive 6. Both result in 2 for the answer.

"Which method did you find easier? The first where we added y to each side, or the second where we subtracted 5y from each side?"

They will likely say they prefer the first because there are fewer steps.

"My preference is ending up with a positive variable term, but after doing these for a while, you will get comfortable with either one."

Variables and Numbers on Both Sides

After they get comfortable doing maybe two more examples, I move on to equations where there are both variables and numbers on both sides. We start with a problem on the board:

$$16 + 4y = 10y - 20$$

Students view the equation as if looking at a car accident on the side of a highway.

One area of confusion that I address immediately is that it doesn't matter whether we work with numbers or variables first.

To reduce distractions, I will sometimes place a Post-It note over one of the terms; let's say it is –20 on the right-hand side so that students now only see this: $16 + 4y = 10y$.

"What do you do now? Subtract $4y$ from each side or 10y from each side?"

The general consensus is usually to subtract $4y$, since that results in a positive variable term. "Let's look at the whole equation now," I'll say, removing the Post-It note:

$$16 + 4y - 4y = 10y - 4y - 20$$

"Someone simplify this for me," I will say.

The result is $16 = 6y - 20$.

"What now?"

They tell me to add 20 to both sides, though I have had students suggest subtracting 16 from both sides. I might say, "But we're so close to getting it where we want it—variable term on one side and number on the other—why would we subtract 16 from both sides?"

We end up with $36=6y$, which they easily solve.

I have had students suggest subtracting $6y$ from both sides instead of $4y$ as I just described. I'll have some students do it that way, to show that we ultimately arrive at the same solution.

Examples. Students work in their notebooks; I offer help and guidance as needed.

1. $4z + 2 = 2z + 8$. *Answer*: $2z = 6$; $z = 3$.
2. $7f - 1 = 29 + f$. *Answer*: $6f = 30$; $f = 5$.
3. $5y + 10 = 6 + 6y$. *Answer*: $4 = y$.
4. $5 - b = b + 5$. *Answer*: $0 = 2b$; or $-2b = 0$; $b = 0$.

I mix simpler equations in so they aren't working on just one type.

5. $7v = 45 + 2v$. *Answer*: $5v = 45$; $v = 9$.
6. $12 - y = 5y$. *Answer*: $12 = 6y$; $y = 2$.

Common Mistake

One mistake I've seen often is subtracting the coefficient from a variable term (e.g., $6y - 6$), in the belief that this equals y. Students need to be reminded why this is not true. The expression $5 \times 3 - 5$ is not 3, but 10. The term $6y$ signifies that 6 is being multiplied by y. To undo multiplication, we divide.

Homework

The homework should be a mix of simple equations where there are variables on both sides, but a number on only one side, and more complex equations with both numbers and variable terms on both sides.

16. PERCENTS AND THE PERCENT EQUATION

Prior to starting the unit on percents. I include more complex equations into the mix in addition to the one I described in the previous chapter. Specifically, I introduce how to eliminate fractions from equations by multiplying all terms by the common denominator.

After the brief foray into more complex equations, I start the unit on percents. One of the goals in teaching this unit is to present an algebraic approach to solving percent problems.

The entire unit on percents is a balance between 1) directly translating a percent problem into an algebraic equation, and 2) using proportions. An example of direct translation is transforming "ninety percent of what number is 22?" into "$0.9x = 22$." The proportion method transforms the sentence into $\frac{90}{100} = \frac{22}{x}$.

Both methods result in the same equation in the end, so I consider both to be algebraic in nature. I have found, however, that most students prefer the proportional method. In the algebra course that they will eventually take, most students become more comfortable with direct translation.

Warm-Ups

1. Solve. $2x - 3 = 4x + 5$. *Answer*: $2x - 2x - 3 - 5 = 4x - 2x + 5 - 5$; $-8 = 2x$; $x = -4$.
2. Write 64% as a fraction in lowest terms. *Answer*: $\frac{64}{100} = \frac{16}{25}$.
3. Solve. $\frac{65}{100} = \frac{130}{x}$. *Answer*: $65x = 13{,}000$; $x = 200$
   or Reduce left – hand side: $\frac{13}{20} = \frac{130}{x}$; $13x = 2{,}600$; $x = 200$.

4. $\frac{16}{25}$ *of* 50 = ? *Answer*: $\frac{16}{25} \cdot \frac{50}{1}$=32.
5. Which fraction is greater? $\frac{5}{7}$ *or* $\frac{4}{5}$? *Answer*: $\frac{5}{7} = \frac{25}{35}$, $\frac{4}{5} = \frac{28}{35}$ *so.* $\frac{4}{5}$ *is greater.*

Problem 1 involves variables on both sides of the equation which they have just had. For Problem 2, students should remember enough about percents that they can do Problem 2. Problem 3 is a proportion which they have learned how to solve. Problem 4 is simple fraction multiplication, which they also have learned but may need a prompt such as expressing 50 as $\frac{50}{1}$ and cancelling.

WHAT IS A PERCENT?

I start off with a quick review of percents. For purposes of review, a basic definition of percent is: A ratio of a number to 100. Fractions can be expressed in terms of 100; for example, $\frac{4}{5}$ equals $\frac{80}{100}$ or 80%.

CONVERTING FRACTIONS TO PERCENTS

We can easily convert $\frac{2}{5}$ to a percent because we can multiply the denominator by 20 to get 100; therefore, $\frac{2}{5}$ is equivalent to $\frac{40}{100}$ or 40%. But we can also divide: $\frac{3}{8} = 3 \div 8 = 0.375$. The quotient is then multiplied by 100 to obtain 37.5%.

TAKING THE PERCENT OF A NUMBER

"What is 90% of 30? Who can tell me?" Usually at least one person does, but sometimes I get blank stares. "What is 90% as a fraction?"

Response: $\frac{90}{100}$.

"Can it be simplified?"

Response: Yes. $\frac{9}{10}$.

"You did Problem 4 of the Warm-Ups; a fraction 'of' a number is the fraction multiplied by the number. So, we can multiply $\frac{9}{10}$ by 30. We can also multiply by the decimal form 0.9, which when multiplied by 30 equals 27. How would I find 47% of 200?" It might take a little give and take with a hint from me, but students soon tell me it's 0.47×200.

Finding the Percent of One Number to Another

"How would I do this? '17 questions out of 25 is what percent?'"

Someone will tell me we can divide, getting 0.68 which is 68%. Or we can multiply numerator and denominator by 4 to get $\frac{68}{100}$, which is 68%.

Putting Sentences into Equation Form

"Finding the percent of a number is fairly easy and one you're used to. But if I asked '19 is 95% of what number?' that's a bit different isn't it? Today's lesson presents a way to use algebra to solve percent problems. There are three types of percent problems: Finding a part of the whole, finding a percent, and finding the whole number."

Finding a Part. "We've been translating English into algebra, like 5 more than a number is $5 + x$. We can use that vocabulary to translate percent problems into very simple algebraic equations."

I write on the board:

"of" : multiply or times.

"is": equals

"what number" or "what percent": x

"For example, in taking a percent of a number like we just did, we can translate it. Suppose I ask, 'What number is 30% of 50?' I know you know how to do this, but I want to show you how to translate it into an equation. First of all, how would I translate 'what number'? When we don't know the number, what do we write?"

Everyone at this point knows it's a variable, and students will shout out "x."

"So 'what number' becomes x. Now, we have 30% *of* 50. First, we have to write 30% as a decimal, which is what?" Response: 0.3.

"How do I write 30% *of* 50; we already know what 'of′ translates to."

They tell me and I write the whole equation on the board: $x = 0.3 \cdot 50$."

Students may ask why they would need to write an equation for that since taking a percent of a number is simply multiplying the decimal form of the percent by the number in question.

"It's just another way for you to solve the problem and to get used to seeing things algebraically." With that, I go on to the two other types of percent problems.

Finding a percent. "The next type is finding a percent of a number. 'What percent of 40 is 6?' is that type of problem. Let's translate it. First, how do I translate 'what percent?'"

Response: x (other letters will do, of course!).

"Now, we have 'of 40.' What does 'of 40' translate to?"

Response: times 40.

"So, we have 'What percent' as x, multiplied by 40. How do I write that?"

Response (inevitably): x times 40".

"And we write that as '$40x$'. Next: How do we write 'is 6'?"

Response: = 6.

"Let's put it all together." I write the equation $40x=6$. Solving it, $x = \frac{6}{40} = 0.15$ or 15%.

Part of the difficulty with these types of problem is the way they are worded. After the initial translation, I reword the problem so it now says: "6 is what percent of 40?" Some students find this easier to translate; others who latch on to the technique of translation say it doesn't matter.

Another issue that sometimes comes up is that an inquisitive student will ask whether "what percent" should be written $\frac{x}{100}$. And indeed it can. If the question comes up, I write the equation using $\frac{x}{100}$ as "what percent." Then, the equation and its solution looks like the following:

$$40\left(\frac{x}{100}\right) = 6;\ \frac{40x}{100} = 6;\ 40x = 600;\ x = \frac{600}{40} = 15$$

It is equivalent to the first method, but "builds in" the multiplication by 100 step. Since $x = 15$, then $\frac{x}{100} = 15\%$. Generally, in my experience, students prefer the first method.

Finding the Whole. "The third type is finding the whole number of which another number is a part. For example: '19 is 95% of what number?' We have '19,' and we have 'is.' How do I write that?"

Response: 19 =

"Now, we have 95%. Do I keep it in percent form?"

Response: No, change it to a decimal.

"So how do we write '0.95 of what number?

Response: $0.95 \times x$.

Which I again correct to $0.95x$.

The equation becomes $19 = 0.95x$.

I ask them to solve it; $x = 20$.

Again, the wording of the original problem may cause difficulty for some. I will ask students if they find this wording easier: "95% of what number is 19?" Some do; others say it doesn't matter.

Examples

1. What number is 10% of 920? *Answer*: $x = 0.1 \cdot 920$; $x = 92$.
2. 120 is what percent of 150? *Answer*: $120 = 150x$ $x = \frac{12}{15} = 0.6$,60%.
3. 15 is 75% of what number? *Answer*: $15 = 0.75x$; $x = \frac{15}{0.75} = 20$.
4. 12 is 40% of what number? *Answer*: $12 = 0.4x$; $x = \frac{12}{0.4}$; $x = 30$.
5. Tom scores 80% on a test with 25 questions how many did he get right?

   "What are we trying to find?" I ask. The response is how many he answered correctly. "Since we don't know that number, what should we call it?" I'm told x. "He had 80% of how many questions?" I'm told 25. "We can say 'What is 80% of 25?" This they now know how to translate:

   $x = 0.8(25) = 20$
6. If Joan had 23 questions right and scored 92%, how many questions were there on the test?

   Again, I work with them to rephrase it: 23 is 92% of what number?

Homework

The homework should be a mix of these type of percentage problems, with students required to write and solve the equations. Mixed in to the problems should be some word problems that involve more than direct translation, but require some interpretation to identify what type of percentage problem it is.

17. SOLVING PERCENT PROBLEMS USING PROPORTIONS

In the previous lesson, students were instructed on how to translate percent problems from English into algebraic equations. In this lesson an alternative is presented that uses proportions. Both methods end up with exactly the same equation.

I have taught students in seventh-grade math who I taught again in eighth grade algebra. When taking algebra these students remarked that the percent problems they found difficult in seventh grade were now very easy. Part of the difficulty they experienced may have been because of their transition to and discomfort in working with equations in seventh grade. More experience with equations in algebra likely led to a better understanding of what is going on mathematically when solving such problems.

In the previous lesson, we discussed three types of computation with percents: 1) finding a part, 2) finding a percent, and 3) finding the whole. In translating these statements directly into equations, students did not have to keep track of what is the part and what is the whole. Using the proportional method it is necessary to know what is what. This is done through a mnemonic method that students gravitate toward.

Warm-Ups

1. In the statement "3 out of 4" what is the part and what is the whole?
   Answer: 3 represents the part and 4 represents the whole.
2. Write as an equation and solve. 15 is what percent of 20?
   Answer: $15 = 20x$; $x = 15/20 = 0.75 = 75\%$.

3. In Problem 2, what number is the part and what number is the whole?
   Answer: 15 *is the part*; 20 *is the whole.*
4. Write as an equation and solve. 16% of what number is 4?
   Answer: $0.16x = 4; x = \frac{4}{0.16} = 25.$
5. Write as an equation and solve. Mary gets 24 questions correct out of 30. What percent of questions did she answer correctly?
   Answer: $24 = 30x; x = \frac{24}{30} = 0.8 = 80\%.$

Many of these problems will elicit questions because students are still unfamiliar with recognizing parts and wholes. For Problem 1, I would hint that if there are 3 slices of pie out of 4 slices can they now identify what numbers are part and whole? Problem 3 may be approached the same way. For Problem 5, a prompt may be: "What are we trying to find? Part, percent, or whole?" It asks for percent so that's fairly easy. Now, it can be restated as "24 is what percent of 30?" or "What percent of 30 is 24?"

Going Over Homework

The previous homework assignment included word problems requiring students simplify it to a part, percent, or whole problem. Usually, problems that require finding the whole prove most difficult for students. I go over these, getting students to first identify whether the problem is asking for part, percent, or whole and then asking them to rephrase the problem in the format for the particular type.

Part/Whole Proportions: Find the Percent

We get into the main part of the lesson with the introduction of the percent proportion, which I write on the board:

$$\frac{part}{whole} = \frac{x}{100}$$

"We can use this proportion to find the solutions to the types of percent problems we've been doing. Let's do one we've done before. '8 is what percent of 20?' In the problem, what are we trying to find? Part, whole or percent?"

It's easy to see it is percent. "The percent value is what *x* represents. We're trying to find what x is in this problem. So the first thing we write is $\frac{x}{100}$. Look at the problem and tell me what the part is and what the whole is."

As needed, I might provide a prompt: "If I said 8 questions correct out of 20 questions, what would be the part and what would be the whole?" Students see that 8 represents the part and 20 represents the whole. "We plug this into our proportion equation."

$$\frac{part}{whole} = \frac{8}{20} = \frac{x}{100}$$

"Now, solve it." They have solved proportion problems using cross multiplication. $20p = 800$; $x = 40$. "So, the answer is 40%."

Some students might find this confusing since if $x = 40$, why is 40 not the answer? I make clear that the problem asks what the percent is, which is $\frac{40}{100}$ or 40%.

I introduce a mnemonic device that helps identify part and whole. "If you have trouble identifying which number is the part, and which is the whole, look at the problem and focus on the words 'is' and 'of.'"

Looking at "8 is what percent of 20?," we find the word "is" next to the number 8, and "of" next to the word 20. "Think of the phrase 'is part of.' The word 'is' connects to the number that is the part. The word 'of' connects to the number that is the whole. So, we can think of the proportion formula as:

$$\frac{part}{whole} = \frac{is}{of} = \frac{x}{100}$$

"Let's try this out. What percent of 30 is 24?" This is worded slightly differently. "What are we trying to find? Percent, part, or whole?"

They tell me percent. "So, we're trying to find x. What's the first thing I write?" Having just completed a problem in the same form, students quickly answer $\frac{x}{100}$.

"This leaves the part and the whole to plug in, which we're calling the 'is' and the 'of.' How do I write that?"

The consensus is $\frac{24}{30}$. I will ask a student to explain why they wrote it as $\frac{24}{30}$. The answer will likely be that "is" is next to 24, and "of" is next to 30.

I move on to the other types of problems, which help solidify both the concept and the mnemonic. It is good to have both, since sometimes identifying what "is" or "of" is connected to can be confusing.

Part/Whole Proportions: Finding a Part:

Problem on the board: What number is 8% of 75?

"If we know the percent, which we do in this case, that's the first thing we write."

$$\frac{8}{100} = \frac{part}{whole}$$

"Now, we have to find the part and the whole. That leaves 'What number' and '75.' So which is part and which is whole?"

I'll go around to see what people write in their notebooks and ask someone who had it correct to tell the class what it is and how the student arrived at it.

The student might say, "The word 'is" is next to 'what number' which is x and 'of" is next to '75.'" Occasionally, they might say that we're finding a number that is a part of the whole, which is 75—but not often in my experience.

I will then point out that the word "is" is also next to 8%. "But we already took care of 8% by writing $\frac{8}{100}$. So that eliminates that possibility."

I ask if someone has the proportion written and have them write it on the board and solve it.

$$\frac{8}{100} = \frac{x}{75};\ 100x = 600;\ x = 6$$

I point out that they could make the calculation easier by simplifying the left-hand side to either $\frac{2}{25}$ or just writing it in decimal form. Looking at the latter:

$$\frac{0.08}{1} = \frac{x}{75};\ x = 0.08(75) = 6$$

Part/Whole Proportions: Finding a Whole

Problem on the board: "39 is 78% of what number?"

"What do we do first?" They know by now to write the fraction form of percent: $\frac{78}{100}$.

"Now, we have to find the part and whole. What are they?"

This one is fairly obvious because of the placement of "is" and "of." The equation becomes $\frac{39}{x} = \frac{78}{100}$, or $\frac{39}{x} = \frac{0.78}{1}$, which they then solve. $x = 50$.

Examples. I give various examples as I did in the previous lesson, which they should work out in their notebooks.

Homework

The homework should be similar to the previous lesson, but students are to use the proportion equation. I make it clear that they should write the proportion for each problem, and that is what I consider "showing work."

18. DISCOUNTS, MARK-UPS, TAXES, AND TIPS

PRIOR CONTENT

Finding the percent change is covered in a lesson previous to this (not included in the book). Percent of change directly relates to and provides a basis for discounts, mark-ups, commissions, taxes and so forth, discussed in this section. It is defined as:

$$\textit{percent of change} = \frac{\textit{amount of change}}{\textit{original amount}}$$

By stating it in this way, it can be taught as "new value minus original value divided by original value" which results in negative values when there is a decrease, and positive values for increases. I start out by expressing everything as either *x* percent increase or x percent decrease where x is always positive. I do this by interpreting the amount of change as a positive difference; e.g., a reduction in price from $10 to 5$ would be termed a $5 decrease. Later, when students are more comfortable with the concept, I introduce the option of using negative values and percents.

PRESENT LESSON

I approach discounts/decreases and mark-ups/increases in two lessons. In this, the first part, I cover the basics in which the sales price after a discount is a two-step process: the amount of discount is calculated first, and then subtracted from the original price. Mark-ups, tips and taxes are also calculated in two parts, but are added to the original amount.

In the second part, I teach the unitary method which can be done by a direct algebraic equation, or by proportion. In the accelerated seventh-grade math course, I include a lesson that extends the unitary method using proportion to allow original amounts to be found, given the after-discount, or after-mark-up prices.

WARM-UPS

1. Write as an equation or proportion and solve. 60 is 75% of what number? *Answer*: $60 = 0.75x$; $x = 80$; or $\frac{75}{100} = \frac{60}{x}$; $75x = 6000$; $x = 80$.
2. Green paint is made with 2 parts blue and 3 parts yellow. If you have 24 gallons of yellow, how much blue do you need, using this ratio? *Answer*: $\frac{2}{3} = \frac{x}{24}$; $x = 16$.
3. Write as an equation or proportion and solve. 18 is what percent of 25? *Answer*: $18= 25x$; $x = 0.72$ *or* 72%; *OR* $\frac{x}{100} = \frac{18}{25}$; $x = 72$, *so* 72%.
4. Write as an equation or proportion. What is 25% of 2000? *Answer*: $x = 0.25(2000)$ or $\frac{25}{100} = \frac{x}{2000}$; $x = 500$.
5. A watch that normally sells for \$2,000 is on sale for \$1,500. What is the percent decrease? *Answer*: $\frac{\textit{change in amount}}{\textit{original amount}} = \frac{500}{2000} = 0.25$ or 25%.

Problems 4 and 5 segue to the portion of the day's lesson on discounts. Problem 4 asks for an equation or proportion but can also be solved by simply multiplying the decimal form of percent by the number; that is, 0.25×2000, as they've done. For calculating discounts and mark-ups, it will be more straightforward.

CALCULATING DISCOUNTS

The discussion of the problems in the Warm-Ups gives us entry into the day's lesson. Specifically, Problem 4 asks what is 25% of 2000. Problem 5 shows that a \$500 reduction in price represents a 25% decrease.

"This is an example of a discount," I say and then ask students if they've seen advertisements that say "25% off". Most everyone has. I then write the following definition on the board:

A discount is the amount by which the regular price of an item is reduced. The sale price=original price – discounted amount.

I write a problem on the board and the steps for solving, which they write in their notebooks:

A bicycle that normally sells for $1,400 is discounted by 20%. What is the discounted selling price?

Step 1: Find amount of decrease by finding the percent of the full price. In this example it is $1400 × 0.2 = $280.

Step 2: Subtract amount of decrease from original amount: $1400-$280 = $1120.

(Again, for purposes of this lesson students should be reminded that finding a percent of a value means multiplying the decimal amount of the percent by the value—it is not necessary to write an equation or proportion though students may do so if they wish.)

Examples:

1. A bicycle normally sells for $200. It is discounted 25%. What is the sales price? *Answer*:$200 × 0.25 = 50; $200 − $50 = $150.
2. Jenny bought a fan that was 30% off the original price of $200. What was the amount of the discount? *Answer: We want the amount by which the price is reduced not the sales price. $200 × 0.3 = $60 reduction.*
3. A sofa priced at $900 is discounted 75%. What is the sales price? *Answer*: $900 × 0.75 =$ 675; $900 − $675 = $225.

Now, let's look at a problem that recalls the previous lesson.

What is the discount rate for a radio that originally sold for $150 and now sells for $120? *Answer*: $\frac{(150-120)}{150} = \frac{30}{150} = 0.2$ *or* 20% discount.

Calculating Mark-Ups, Tips, and Taxes

Students know what tips and taxes are. They may not know what a mark-up is, so I explain how that works. The main point is that the final value of a marked-up item, or the cost of something with tips or taxes, are all calculated in the same way.

"For discounts, we took the value of the reduction and subtracted it from the original price. How do you think we calculate the price of a \$100 meal with a 15% tip added?"

Students make the connection that if discounts involved subtraction, then mark-ups, tips and taxes involve addition. I write on the board and then work out an example:

A mark-up is the amount by which the regular price of an item is increased. The sale price = original price + increased amount.

Example of \$100 meal with 15% tip.

Step 1: Find the amount of increase by finding the percent of the original price. (\$100 × 0.15 = \$15).

Step 2: Add the amount of increase to original price. (100 + 15 = \$115).

EXAMPLES

1. 15% tip on \$35 restaurant bill. What is the amount of the tip?
   Answer: 0.15 × 35 = \$5.25.
2. What would be the total price for a \$1200 bicycle with 8% sales tax? *Answer: \$1,296.00.*
3. A 2 lb can of coffee costs \$12.00 at the Star Market. If the manager buys the coffee for \$8.00 a can, what percent markup does he allow on each can? *Answer*: $\frac{12-8}{8}$ = 0.5 *or* 50% *markup.*

Mix of discount and tax: Suppose a calculator costs \$75 originally and is discounted by 20%. What is the final price if the sales tax is 6%?

First, I ask students to find the discounted price: \$75 × 0.2 = 15; \$75 − \$15 = \$60.

Now, the tax is calculated and added on: \$60 × 0.08 = \$4.80; \$60 + \$4.80 = \$64.80.

HOMEWORK

In my experience students find these problems straightforward. Homework should be a mix of discount and mark-up, tax, and tip problems, some asking for the discounted amount, some asking for the marked-up amount. Also, there should be problems that require finding the amount by which an item is reduced, as well as problems asking for the amount of a mark-up. Problems requiring finding what the discount, tip, tax, or mark-up rates are (using percent of change) should also be included.

19. UNITARY METHOD FOR DISCOUNTS AND MARK-UPS

The current lesson focuses on the "unitary method" for discounts—a method that allows for the calculation to be done in one step rather than two. It also provides the foundation of an algebraic approach that will be built upon in a subsequent algebra course. The unitary method is presented in two ways. The first way is direct translation to an algebraic equation as has been introduced and practiced in a previous lesson on percent problems. And, as was done before, I present a proportional method as well, to which most seventh-graders gravitate.

Some students will cling to the two-step method. In the accelerated seventh-grade math course, students learn how to find the original amount when the discounted (or marked-up) value is given. I've found that students in that course also rely on the proportional method while a select few feel comfortable using a direct algebraic translation approach.

Subsequent Lesson and Placement

The subsequent lesson addresses the unitary method for mark-ups and other increases but is omitted from this book. It follows the same pattern as the unitary approach for discounts except that the percent increase (as a decimal) is added to rather than subtracted from one, to serve as the multiplier.

The lessons on the unitary approach are normally taught in accelerated seventh-grade math courses. I include the unitary approach in this section of the book because I have taught it in both courses due to its close relation to the two-step method.

Warm-Ups

1. A bookshop gives a 20% discount during a sale. Tom bought two books which cost \$40 and \$15. How much did he have to pay altogether?

   Answer: $65 - (65 \cdot 0.2) = 65 - 13 = \$52$.
2. What is the discount rate on a bottle of shampoo that is reduced from \$20 to \$12? Answer: $\frac{20-12}{20} = \frac{8}{20} = 0.4$ *or* 40%.
3. A \$600 television is discounted 24%. What is the amount of the discount? *Answer*: $0.24 \times 600 = \$144$.
4. Susan and Nancy received some money in the ratio of 2:5. Nancy received \$36 more than Susan. How much money did Susan receive? (Use algebra or tape diagram to solve.) *Answer*: \$24: *Let* $2x$ = *amount of Susans money*, $5x$ = *amount of Nancys money*; $5x - 2x = 36$; $3x = 36$; $x = 12.2x = 24$, $5x = 60$.
5. John's uncle says he will help John purchase a bike by paying 30% of the cost. If the bike costs \$800, what will John's share of the cost be? *Answer*: *John will pay* 70% *of the cost*; $0.7 \times 800 = \$560$.

Problems 1–3 are discount problems that track what students have done in the previous lesson. Problem 5 is directly related to this lesson. One prompt I use for students who ask for help is "What if his uncle paid 50%; what would be John's share?" Obviously 50%. Then, "What if he paid 60%, what's John's share?" They see it is 40% and that the uncle's share is subtracted from 100%. This reasoning is used to solve Problem 5.

Finding Discounts Using the Unitary Method 1

Since we have gone over Problem 5 of the Warm-Ups, students know that John's uncle chipped in 30% and that John's share is figured to be 100% – 30% or 70%.

"Let's say there's a \$900 bike on sale 25% off. What is your share going to be?"

Hearing "75%" I continue: "How much are you going to pay? Can we write it as an equation?" I ask students to work on mini-whiteboards or in their notebooks, and I go around to see that they are doing it correctly and getting \$675."

"This is a different way of finding the discounted price. It's called the Unitary Method, because instead of doing the calculation in two steps like we did yesterday it's done in one step. There's no subtraction step."

(Actually there is a subtraction step—the discount rate is subtracted from 100%. But in all the time I've been teaching this, no student has ever pointed that out.)

Examples

I ask students to work with me on a problem, as a worked example: "A radio normally sells for \$140. It is discounted 20%. What is the sales price?"

I then ask: "What percent are you going to pay?"

Hearing 80%, I state, "This is now asking '80% of 140 is what number'? How do we write that as an equation?"

I have them write on their mini-whiteboards or notebook. It should be $x = 0.8 \cdot 140$. (Typical mistakes are not changing the 80% to 0.8.) The answer is \$112.

1. A \$420 watch is discounted 15%. What is the sales price? *Answer*: $0.85 \cdot 420 = \$357$
2. Mr. Hallmeyer pays 70% of the price for a necklace, which normally sells for \$400. What is the discount rate? *Answer: 30%.*
3. What is the amount of the discount for Problem 2? *Answer*: $x = 0.3 \cdot 420 = \$126$.

This problem asks for the amount of the discount, not the final sales price.

Finding Discounts Using Unitary Method 2

In the unitary method described above, some students forget to subtract from 100; others find the method confusing. "There is another method that we'll now learn that uses the proportion method. Some of you may find this method easier."

The proportion method uses this formula:

$$\frac{100 - \textit{discounted rate}}{100} = \frac{\textit{discounted price}}{\textit{original price}}$$

To show how it works I give an example: "A $60 book is discounted 15%; find the sales price."

"Let's plug in the numbers to the formula. The discount rate is 15%, so we plug in 15. Now let's work on the right-hand side. What are we trying to find out?" I want to hear we are trying to find the discounted amount.

"And if we don't know what the value is, what do we use?"

Response: x.

"Do we know the original price? What is it?"

Response: $60.

The proportion now looks like this: $\frac{100-15}{100} = \frac{x}{60}$. This simplifies to $\frac{85}{100} = \frac{x}{60}$.

"You can make this a bit easier on yourselves by simplifying the left-hand side. Can anyone tell me how we write the left-hand side as a decimal?" Hearing "0.85," I write: $\frac{0.85}{1} = \frac{x}{60}$. "Now, we cross multiply and solve." $x = 0.85 \cdot 60 = \$51$.

"This looks like the form we get when we use the first method," I say. The "100 – discount rate" in the numerator forces students to do the subtraction that proves elusive for some.

EXAMPLES

1. A $150 CD player is discounted 18%. What is the sales price? *Answer*: $\frac{100-18}{100} = \frac{x}{150}$; $\frac{82}{100} = \frac{x}{150}$; $\frac{0.82}{1} = \frac{x}{150}$; $x = 0.82 \cdot 150$; $x = \$123$.
2. A school had 2,000 students, but this year that number was reduced by 22%. What is the current number of students? *Answer*: $\frac{100-22}{100} = \frac{x}{200}$; $\frac{0.78}{1} = \frac{x}{200}$; $x = 0.78 \cdot 2000 = 1{,}560$ students.

This problem is not a discounted price problem, but the structure and method is exactly the same. A prompt I use: "Population is reduced by a 22% decrease. Set it up as if the population were money."

3. A laptop sells for $1,750. It is discounted by 45%. By how much is the price reduced? *Answer*: $x = 0.45 \cdot 1750$; $x = \$787.50$.

Before students dive into the problem, I ask, "What question does the problem want answered? Final price or amount of discount?" In fact, it is asking, "What number is 45% of 1,750?"

Homework

Homework problems consist of problems that involve discounts or other types of reductions. Students are free to solve using whatever method they find easier. Some problems will ask for the amount of a discount or reduction, in which case students simply multiply the percent reduction by the original amount. Mixing these problems is important to reinforce the skill of reading a problem and identifying what it is asking.

20. FINDING ORIGINAL AMOUNTS

In previous lessons, two proportions were introduced that allow finding both discounted and marked-up values:

$$\textit{Decreases}: \frac{100 - \textit{percent decrease}}{100} = \frac{\textit{reduced amount}}{\textit{original amount}}$$

$$\textit{Increases}: \frac{100 + \textit{percent rate of increase}}{100} = \frac{\textit{increased amount}}{\textit{original amount}}$$

In this lesson, the focus is on finding the original amount given 1) the discounted or marked up value and 2) the percent decrease or increase. Example problem: "John received a 20% raise and now receives $1,200 per week. How much did he make before the raise?"

The method uses the same proportion equations students used to find the final discounted or marked-up values. The challenge to students is to identify the type of problem (e.g., find discounted amount or find original amount) and keep track of information so they know where the values are placed in the proportion equation.

Although these problems can also be solved by translating directly into equations, my experience is that seventh-graders find the proportion equation easier to use. The above problem, solved by direct translation to equation, would be to let x equal the salary prior to the raise, and adding 20% of that amount to it to represent the new salary: $1.2x = 1200$; $x = \$1,100$. As with finding discounted and marked-up amounts, the proportion method used for finding the original amount leads to the same equation obtained using direct translation.

Warm-Ups

1. The ratio of John's weight to Peter's weight is 5:3. Their average weight is 40 lb. Find John's weight. (Hint: What is their total weight? Use tape diagrams or algebra.) *Answer*: 50 *lbs. Let* $5x$ = *John's weight*, $3x$ = *Peter's weight. Total weight* = 40 × 2 = 80. *Then*, $5x + 3x = 80$; $8x = 80$; $x = 10.5x = 50$, $3x = 30$.
2. Mr. Jones' salary of $1,000/week increased by 25%. The next week because of cut-backs, his company reduced his salary back to $1,000. What was the percent of decrease? (It's best to show your work on this one.)

   Answer: *Increased salary*: $\frac{100+25}{100} = \frac{125}{100} = \frac{x}{1000}$; $\frac{1.25}{1} = \frac{x}{1000}$; $x = 1{,}250$.

   Percent of decrease: $\frac{1250-1000}{1250} = \frac{250}{1250} = 0.2$ *or* 20%.
3. There are 40 people at a party at 7 p.m. By 8 p.m., that number has increased by 25%. How many people are at the party at 8 p.m?

   Answer: $1.25 \cdot 40 = 50$. *OR*: $\frac{125}{100} = \frac{x}{40}$; $\frac{1.25}{1} = \frac{x}{40}$; $x = 1.25 \cdot 40$; $x = 50$.
4. A watch originally selling for $1,000 is discounted to $800. What is the rate of discount? (Hint: This is asking for the rate of change)

   Answer: $\frac{1000-800}{1000} = \frac{200}{1000} = 20\%$ discount.
5. 85% of what number is 51? *Answer*: $0.85x = 51$; $x = \frac{51}{0.85}$; $x = 60$ *or* $\frac{85}{100} = \frac{51}{x}$; $\frac{0.85}{1} = \frac{51}{x}$; $0.85x = 51$; $x = 60$,

Problem 1 is a review of ratio problems that are solved using either tape diagrams or algebraic equations. Problem 2 requires the new salary to be calculated; the percent decrease is then calculated using the percent of change proportion. Many students think the answer will be 25% decrease. The remaining problems focus on various aspects of percents. Problems 3 is a percent increase problem. Problem 4 is a percent of change problem. Problem 5 requires students to solve using either an equation or the proportion method.

Finding Original Amounts Using Proportions

I start the class off with a re-cap of what we've done the past two lessons. "We learned how to solve decrease and increase problems using the unitary method. And just to make sure you remember what we did, let's do a problem. Suppose there is a jacket that sells for \$60 and it is discounted 15%. How do I find the sale price?"

They work it in their notebooks. I pick two students, one who used an algebraic approach and one who used the proportion approach, and ask both to put their worked solutions on the board.

Algebraic approach: $x = 0.85 \cdot 60$; $x = \$51$.

Proportion approach: $\frac{100-15}{100} = \frac{x}{60}$; $\frac{80}{100} = \frac{x}{60}$; $\frac{0.85}{1} = \frac{x}{60}$; $x = 0.85 \cdot 60 = \$51$.

"Now, let's pretend we don't know that the original price is \$60, but we only know that the sales price is \$51 and that the discount is 15%. And now let's set up the proportion equation that we've been using for discounts."

$$\frac{100 - \textit{percent decrease}}{100} = \frac{\textit{reduced amount}}{\textit{original amount}}$$

"We know the discount rate is 15%, so we plug that in."

$$\frac{100-15}{100} = \frac{85}{100} = \frac{\textit{reduced amount}}{\textit{original amount}}$$

"Now, I want to put the 51 somewhere on the right-hand side. Where do I put the 51? Is 51 the original amount or the reduced amount? Write it in your notebooks."

Checking, I look to see that 51 is placed in the numerator, signifying reduced amount:

$$\frac{85}{100} = \frac{51}{x} \ \textit{or} \ \frac{0.85}{1} = \frac{51}{x}$$

Solving it, we obtain $0.85x = 51$, *and* $x = \$60$.

"And so we've found the original amount. Let's try another. A dealer buys a CD player and marks it up 60% and sells it for \$320. What did the dealer pay for it? Can we solve this in similar way?"

As a prompt, I will ask for the proportion equation for mark-ups. They will know this, and if they have forgotten from the last lesson, I tell them to look it up in their notes. I ask a student to write it on the board:

$$\frac{100 + \textit{percent increase}}{100} = \frac{\textit{increased amount}}{\textit{original amount}}$$

Again, I ask questions to get them to put numbers and the variable in their proper places in the equation, ending up with:

$$\frac{100{+}60}{100} = \frac{320}{x};\ or\ \frac{160}{100} = \frac{320}{x};\ \frac{1.6}{1} = \frac{320}{x};\ 1.6x = 320;\ x = \$200$$

Now, I proceed to the examples which include both discounts/decreases and mark-ups/increases. Students usually see the pattern but need practice identifying the values and placing them properly.

Examples

1. A fan is on sale for 30% off at \$140. What was its original price?

Prompts: "Do we add 30% to 100% or subtract it from 100%. Where does the 140 go: numerator or denominator?" The proportion equation is now:

$$\frac{100-30}{100} = \frac{140}{x};\ \frac{0.7}{1} = \frac{140}{x};\ 0.7x = 140;\ x = \$200$$

2. Mr. Tyler received a raise of 10% and is now making \$1,408 per week. How much did he make before the raise?" *Answer*: $\frac{110}{100} = \frac{1408}{x}$; $1.1x = 1408$; $x = \$1,280$.
3. Another fan is on sale for 30%. The original price was \$300. What was the sale price?

   I throw in this problem for good measure so students know how to go back and forth using the same proportion equation:

$$\frac{100-30}{100} = \frac{x}{300};\ \frac{70}{100} = \frac{x}{300};\ \frac{0.7}{1} = \frac{x}{300};\ x = 0.7 \cdot 300;\ x = \$210$$

More Examples

These include finding original value as well as reduced or increased value.

4. Juan sold a bicycle at a discount of 15%. The sales price was $340; find the original price. *Answer*: $\frac{0.85}{1} = \frac{340}{x}$; $0.85x = 340$; $x = \$400$.
5. Jake spent 10% more this week than last week. He spent $55 this week. How much did he spend last week? *Answer*: $\frac{1.1}{1} = \frac{55}{x}$; $1.1x = 55$; $x = \$50$.
6. Joan bought a blouse that was 35% off. The original price was $60. What is the sales price? *Answer*: $\frac{100-35}{100} = \frac{0.65}{1} = \frac{x}{60}$ $x = 0.65 \cdot 60$; $x = \$39$.
7. Tyler increased his test score by 10%. He previously scored 80. What is his new score? Answer: $\frac{110}{100} = \frac{x}{80}$; $x = \frac{1.1}{1} = \frac{x}{80}$; $x = 1.1 \cdot 80 = 88$.

Finding Original Amounts Using Equations

As necessary, I will do this part of the lesson the next day so as not to overload students. I start by giving a discount problem: "A radio discounted 25% sells for $120. What was its original price?

"If it's discounted 25% what percent do you pay?"

Having done problems using the proportion formula, more students are now internalizing the subtraction step, and I soon hear: "75%"

"So, 75% of the original price is $120. How would you translate that?"

I ask students to write in their notebooks and check their progress. They should have $0.75x = 120$; $x = \$160$. For comparison purposes, I ask them to now solve it using proportions; they will get the same equation.

Examples

1. John made $44 an hour after receiving a 10% raise. How much did he make before the raise?

   Answer: *Let x = salary before the raise.* $1.1x = 44$; $x = \frac{44}{1.1}$; $x = \$40$.

Prompts: "What do we add 10% to?" "110% of what amount is 44?" "Can you translate that into an equation?"

2. A tire is sold at a discount of 10%. It is sold for $45. Find the original price of the tire. *Answer*: $0.9x = 45$; $x = \frac{45}{0.9}$; $x = \$50$.

4. If a bicycle sells for \$1,500 and is discounted 30%, what is the sales price of the bicycle? *Answer*: $x = 0.7 \cdot 1500 = \$1050$.

HOMEWORK

The homework should start off with finding original amounts and progress to problems that require finding original and discounted or increased amounts. Interspersed should be problems in which the amount of the increase or decrease is calculated. For example: "A suit sells for \$500, and is discounted 30%. What is the amount of the reduction?" *Answer*: $x = 0.3 \cdot 500$; $x = \$150$ or $\frac{30}{100} = \frac{x}{500}$; $\frac{0.3}{1} = \frac{x}{500}$; $x = \$150$.

PART II.B
SAMPLE OF TOPICS FROM ACCELERATED SEVENTH GRADE AND MATH 8

So far, discussion of seventh-grade math has focused on regular Math 7 classes, with some blending of accelerated Math 7 topics. Those topics may or may not be doable and are dependent on the ability of the students in said classes.

At this juncture, we now depart from the regular Math 7 curriculum and address the accelerated Math 7 program. Because those topics are the same as those addressed in Math 8, this chapter covers both grades. Geometry topics are not addressed in this book as mentioned in the introduction.

The topics covered are as follows:

21. Linear Equations and Functions

22. Representing Linear Functions

23. Slope and Constant Rate of Change

24. Slope-Intercept Form of Linear Equations, and Graphing

25. Systems of Equations: Solving by Graphing

26. Systems of Equations: Solving by Elimination

21. LINEAR EQUATIONS AND FUNCTIONS

Functions are an abstract concept that at the novice level must be taught in a basic form that is easily understood: linear equations. As students progress through subsequent math courses, the definition essentially stays the same, except there are more functions introduced: quadratic, cubic, trigonometric, exponential, logarithmic, and so on.

Functions are most easily described and defined in terms of values that are dependent on one another. For example, the cost of a bag of oranges that sell for $3 per pound is dependent on how many pounds are bought. Functions are introduced as rules that assign a number to another one. At this level, the rules are in the form of equations.

Typically, textbooks introduce relations first, as ordered pairs of numbers, and then introduce functions as a special type of relation in which for each "input value" there is one and only one "output value." These textbooks then ask students to find examples of relations that aren't functions, and those that are, by virtue of whether each input has one output.

I introduce the difference between relations and functions in a subsequent lesson and then turn to graphing. At that point, having digested the initial concept of what a function is, students are better prepared to build on that foundation.

Warm-Ups

1. Solve the equation. $3y - 12 = y - 2$. *Answer*: $2y = 10$; $y = 5$.
2. In the equation. $y = -4x + 2$, what is the value of y if $x = 3$?
   Answer: $y = -4(3) + 2 = -12 + 2 = -10$.

3. A rental company charges \$50 for a moving truck plus an additional fee of \$0.55 per mile that the truck is driven. If the truck drives 100 miles, what is the total cost for the rental? *Answer*: 50 + (0.55)(100) = 50 + 55 = \$105.
4. An egg tray with 12 eggs weighs 440 grams. The empty tray weighs 20 grams. What is the average weight of an egg? *Answer*: *Weight of 12 eggs*: 440 − 20 = 420; *average weight* = $\frac{420}{12}$ = 35 *g*.
5. A bowling alley charges a \$5 fee for renting shoes and \$2 per game. If Jerry pays \$25 for bowling, how many games did he bowl? *Answer*: *Let x = number of games bowled, so* $2x$ = *cost for x games*; 25 = 5 + $2x$; $2x = 20$; $x = 10$ *games.*

With the exception of Problems 1 and 4, the Warm-Ups focus on how the value of one variable affects another; that is, the relationship between independent and dependent variables. This section addresses such dependencies in terms of what functions are and how they work.

What We Know So Far

The discussion of Warm-Up answers leads to the topic of functions. "In the Warm-Ups, you may notice that in Problem 1, there is only one variable. But in Problem 2, there are two variables. If we know what x equals, we can find out what y equals."

"Now, we'll look at Problem 5 about the bowling alley. In that problem, you solved an equation given that the total cost of the games was \$25. Now, let's look at it without knowing the actual cost. So, there will be two variables; number of games, and total cost of x games."

I write the following equation on the board: $y = 5 + 2x$. I ask the following questions and have indicated what the desired responses are.

"From Problem 5, what does the 5 represent?" *The cost of shoe rental.*

"What does x represent?" *The number of games.*

"What does the 2 in $2x$ represent?" *The cost of each game; \$2.*

"So y would be what?" *The total cost of x games.*

"The value of y changes depending on what x is; that is, the number of games. If the number of games increases, what happens to y?"

I hope to hear "y increases," but I'll accept "It costs more," and then paraphrase.

"If we have an equation $y = 2x$, the value of y depends on what x equals. If x is the number of pounds of oranges, and oranges cost $2 per pound, then $2x$ is the cost of x pounds of oranges? If we bought 2 pounds one week and 3 pounds the next, the cost y would change from what to what?"

Response: From $4 to $6.

"The cost depends upon the price, because the values of y change as the values of x change. If x decreases, what happens to y?

Response: It decreases.

What Is a Function?

"Instead of saying 'the cost depends upon the price,' we can also say that 'the cost is a function of the price' or 'y is a function of x.'"

"For example, let's say we have $y = 2x$. This rule says for each value for x, say 1, 2, 3, 4, y is twice the amount. What are the output values; that is, what does y equal?,

Response: 2, 4, 6, 8.

"If you travel at 40 mph, and the time of travel is t hours, then the distance d depends on the value of t. And we can express this relationship as $d = 40t$."

I ask students to write in their notebooks:

A function is a rule that assigns to each input value one and only one output value.

(The "one and only one output value" is covered in the subsequent lesson.)

The "input values" are called the "domain," and the "output values" are called the "range."

"We saw that when we have an equation like $d = 40t$, as t changes, so does d. The d value is an output value; t is an input value. In this equation, d is dependent on t. So, which value, d or t, is the dependent value?"

Response: d.

"So, the independent variable is *t*; it can be whatever we want it to be. It's free as a bird. But *d* is dependent on *t*."

Examples. The examples focus on expressing functions in various forms.

"Functions can be in the form of an equation, a list of ordered pairs, a table, or a graph," I say. "So, let's look at this more closely."

1. For $y = 3x$, the y values are dependent on the *x* values. Whatever *x* equals affects what *y* equals. If the *x* values are 1, 2, 3, 4, 5, what are the associated *y* values?"

   Answer: 3, 6, 9, 12, 15.

"List these as ordered pairs. I'll do the first one; you do the rest. (1,3). What would we get if *x* is zero?" (0,0).

"Now, do 2, 3, 4, and 5." *Answer*: (2,6),(3,9),(4,12),(5,15).

"Now, let's write this as a table. We'll put the *x* values in the top row and *y* values in the second row. Write in your notebook and fill in the *y* values."

x	1	2	3	4	5
y					

I circulate around to see that they have done this correctly:

x	1	2	3	4	5
y	**3**	**6**	**9**	**12**	**15**

2. For the equation $y = 2x + 3$, I ask students to write the ordered pairs, and then draw a table using 0, 1, 2, 3 and 4 for the *x* values. *Answer*: (0,3)(1,5)(2,7)(3,9)(4,11).

x	0	1	2	3	4
y	**3**	**5**	7	**9**	**11**

"What is the domain?" (If they've forgotten, I remind them they're the input values) 0, 1, 2, 3, 4 "What is the range?" 3,5,7,9,11.

"What's dependent on what? Which is the dependent value, and which is the independent value?" *Answer: y is dependent; x is independent.*

"The domain values are independent; the range values are dependent." I ask students to write that statement down in their notebooks.

3. A boat travels at 5.5 mph. The distance it travels is a function of the time traveled. If it travels for 2 hours at 5.5 mph, how far has it travelled? For 4 hours? *Answer*: 11 *miles*.4 *hours*? 22 *miles*.

Homework. The homework should be problems like the examples. Write ordered pairs and a table for various functions. Identify functions, as in Example 3. Identify the domain and range, and the dependent and independent variables.

22. REPRESENTING LINEAR FUNCTIONS

PRIOR CONTENT

Related to the definition of function is "relation," which is defined as a set of ordered pairs of numbers. If a relation has ordered pairs in which no two pairs have the same first number and different second numbers, then that relation is a function. If a relation has two or more ordered pairs of numbers in which the first numbers are the same but the second numbers are different, then it is not a function.

An example of a relation that is not a function is this set of ordered pairs. (36,6),(36,−6),(25,5),(25,−5),(30,5),(31,5) For these ordered pairs, there are some pairs for which the x coordinate has two different y coordinates; for example, (36,6) and (36,−6).

An example of a relation that qualifies as a function is the following set: (0,0),(1,1),(2,2),(3,3),(4,4). There is one and only one output for each input value.

CURRENT LESSON

In this section, we show how functions can be represented graphically. I've found that it helps to show that an equation such as $16 = 5x + 1$ has one answer that satisfies $y = 5x + 1$. That is, when x equals 3, then $y = 16$. In fact, there are infinitely many such (x,y) combinations for the equation, which is the main message of this lesson.

In most textbooks, the fact that an infinity of ordered pairs are contained on the same line is used as a means for graphing the equation. Students are directed to find three ordered pairs, plot them, and then connect them in a straight line. A subsequent lesson provides a method of graphing such equations by finding the x and y intercepts. (This method is not included in this book.) This method works well for equations in standard form: $Ax + By = C$, where C does not equal 0.

WARM-UPS

1. Translate into an equation and solve. Three more than two times some number is 10. *Answer*: $3 + 2x = 11$; $2x = 8$; $x = 4$.
2. In the equation $y = 2x + 7$ find what y equals when x equals 5. *Answer*: $y = 10 + 7 = 17$.
3. Susan and Nancy received some money in the ratio of 2:5. Nancy received \$36 more than Susan. How much money did Susan receive? (Use algebra or a tape diagram to solve.) *Answer*: \$24; $5x$ = *amount of Nancy's money*; $2x$ = *amount of Susan's money*; $5x - 2x = 36$; $3x = 36$; $x = 12.5x = \$60$, $2x = \$24$.
4. A motorcyclist rode for 7 hours from Smallville to Metropolis at an average speed of 35 mph. A van took only 5 hours for the same trip. Find the speed of the van. (*Hint*: distance = speed × time; speed = distance ÷ time.)

   Answer: *Distance from Smallville to Metropolis is* 7 × 35 = 245 *miles.*

   Speed of van = 245 ÷ 5 = 49 *mph.*
5. 32-45 = ? *Answer*: −13.

These are mostly review problems, although Problem 4 poses difficulty for students. The hints given in the problem are what I point to when students ask how to solve the problem. Other prompts: "What's the distance from Smallville to Metropolis? How do you find it? Once you know it, how can you use it to find the van's speed?" Problem 5 is included because in my experience, seventh-graders forget the rules for working with negative numbers; the more such problems are repeated, the better. Problem 1 relates directly to this lesson.

Linear Functions and Equations

"Let's start by looking at Problem 1 of the Warm-Ups. It translates to the equation $3 + 2x = 11$. There is one answer to the equation: $x = 4$. But if we replace the 11 with y, we get $3 + 2x = y$, which we can write $y = 3 + 2x$. Is there only one answer to the equation?"

Some might say yes and argue that for one value for y, you will get one answer for x. This is true, and I say so. But it is not the point I'm trying to make. "If you can put in other values for x, you will get more answers for y; in fact, there are many answers to this equation. What does y equal if x equals 3?"

The answer is 9, and I write (3, 9) on the board. I throw out some more numbers for students to plug in for x and write down more ordered pairs: (2,7),(0,3) (-2,-1).

"The equation $y = 3 + 2x$ is like a rule that assigns numbers in the domain (that is, the input values, in this case x) to numbers in the range—the output values or y in this case. There is only one output value for each input, so it's a function."

Graphing Linear Equations

I ask students to write the following in their notebooks:

A linear function is a function whose graph is a nonvertical line.

"Let's see what this means. We just came up with four ordered pairs for the equation $y = 3 + 2x$, so let's graph it."

I will project graphing coordinates on the whiteboard or alternatively have a poster-sized piece of such a grid, and I ask a student or students to graph the four points:

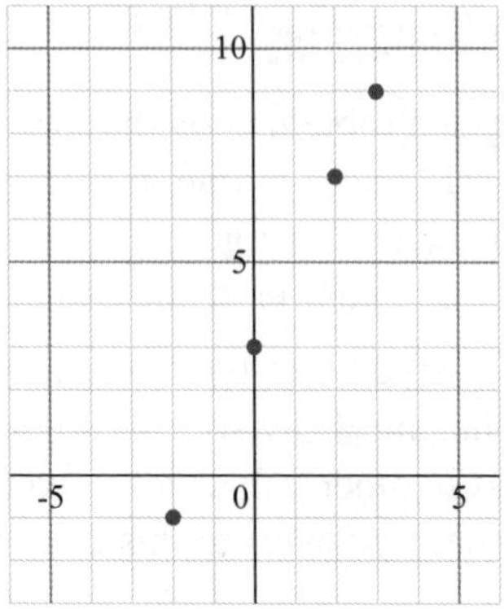

After connecting the points, we get the following graph:

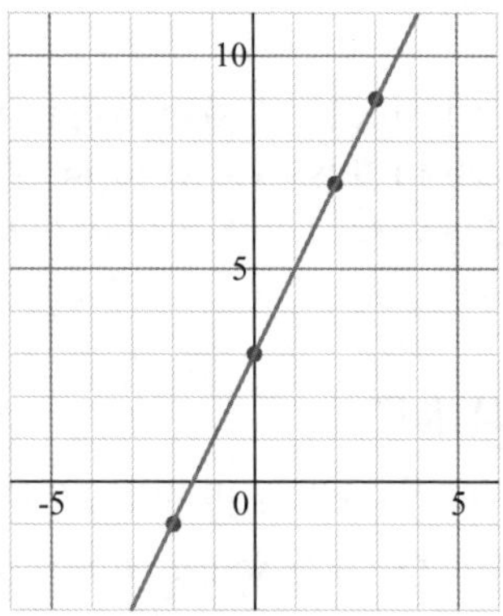

I ask students to write the following in their notebooks:

Any ordered pair on the graph of a linear function is a solution of the related equation.

"So, in this case, if we pick any point on the line, it will be a solution of $y = 3 + 2x$. The line represents every solution to that equation."

The above technique is used as an initial method for graphing linear equations.

"We can use this method to graph an equation. You don't have to use four points; you can use three and still get good results. Let's do one. $y = 5x - 2$."

I ask students to construct a table as follows:

x	$y = 5x - 2$	y	(x,y)
−1			
0			
1			

"To make it easy on yourself, use the numbers −1, 0, and 1, substitute them for x in the equation to get y, and then write the ordered pair." We work through the equation and end up with:

x	$y = 5x - 2$	y	(x,y)
−1	$y = -5 - 2$	−7	(−1, −7)
0	$y = 0 - 2$	−2	(0, −2)
1	$y = 5 - 2$	3	(1, 3)

Students then plot the points in their notebooks and connect them with a straight edge:

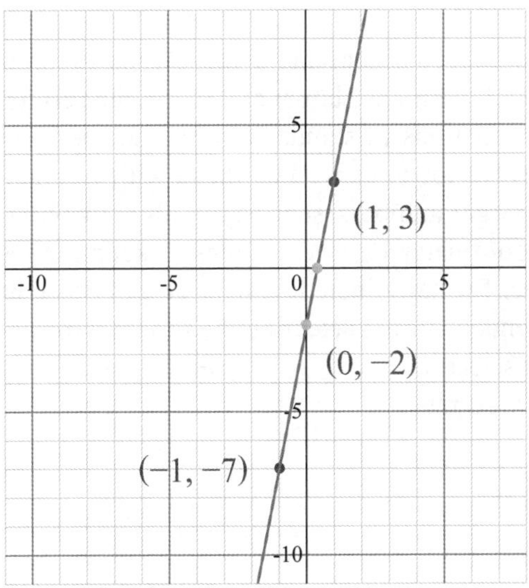

Examples

I continue with two similar problems and then give the following: $3x + y = 5$.

The key here is to solve for y, to obtain $y = -3x + 5$, which they can then work as before. Students have difficulty at first with such problems. These problems are limited to equations where the coefficient of y is 1. This then requires only that students move the x term to the other side.

Next, I give an equation that has a fractional coefficient. $y = \frac{2}{3}x - 2$.

I usually have a few students who still have difficulty multiplying a whole number by a fraction.

"To make this easy, pick values for x that are –3, 0, and 3 so that the three's cancel."

The ordered pairs obtained will be (–3,–4),(0,–2), and (3,0).

Homework

The homework is a continuation of these type of problems; students construct tables of values as shown and graph the results. The problems should include equations in standard form, such as $2x + y = 3$, so that students must solve for y, as well as those with fractional coefficients. Equations with larger numbers, such as $y = 20x + 10$, will require students to label their coordinates by fives or tens to keep the resulting graph a manageable size. I usually include one or two of those in the examples, telling the students that I am selecting the problem from their homework assignment, thus leading them to believe they are getting the answers "for free." (They aren't!)

23. SLOPE AND CONSTANT RATE OF CHANGE

Previous lessons provided methods for graphing equations. One used a table of values to find three points that were then plotted to determine the line. The other was used for equations in standard form (i.e., $Ax + By = C$). The x and y intercepts are found by setting x and y to 0, respectively, and solving for the other variable. The two intercepts are then connected by a straight line to produce the graph. Students have also learned the $f(x)$ notation for functions.

In this lesson, we look at what slope is, and that it represents a constant rate of change, whether it be distance over time, cost of fruit per pound, and so on. The initial introduction of slope provides the foundation for the concept of rate of change, which continues to be built upon in subsequent math courses. For some students, the concept will start to sink in more during algebra. What students take with them is how the slope of a line is calculated. In a subsequent lesson, they learn how slope is used to find the equation of a line.

Warm-Ups

1. The function $f(t) = 40t$ represents the distance traveled at 40 mph for t hours. What is the independent variable in this function? *Answer: t is the independent variable.*
2. What are the x and y intercepts for the equation $y - 2x = -2$?

   Answer: x intercept: $-2x = -2$; $x = 1$; $(1,0)$ *y intercept*: $y = -2$; $(0,-2)$.

3. Maria has averaged 88 on her last three tests. What does she need to average on her next two tests to have an overall average of 90? *Answer: Total score for last three tests: 3 × 88 =264. Total score for all five tests: 90×5 = 450. Score needed on next two tests to reach 450: 450 – 264 = 186. Average of the last two tests:* $\frac{186}{2}$ = 93.
4. Complete the table of values for $y = 2x$.

x	1	2	3	4
y	*2*	*3*	*4*	*8*

5. Complete the table of values for y =3x.

x	1	2	3	4
y	*3*	*6*	*9*	*12*

Problem 3 will be the most challenging, and students will likely need hints and guidance for solving it. Prompts include: "What would be the total points for the three tests? How do you figure that out? How about for all five tests? How do those totals help you?" Problems 4 and 5 of the Warm-Ups are a direct lead-in to the day's lesson.

RATE OF CHANGE

"In Problem 4 of the Warm-Ups, suppose for the first table, the function represents a remote control car that travels 2 inches per second. And for the second table, we have a remote control car that travels 3 inches per second. The x values represent seconds, and the y values represent inches. For the first car, how many inches does it travel in the first second?"

The answer, of course, is 2 inches.

"How many for two seconds?"

The answer is 4 inches.

"Looking at the first table, how much does the distance increase for each second?" The answer is 2 inches.

"Now, let's find the change in distance from 2 seconds to 4 seconds. What is that change?" The change is 8 – 4, or 4 inches. The time elapsed is two seconds. "What is the ratio?"

I should hear $\frac{4}{2}$, which simplifies to 2 in/sec.

I do the same thing for the second table in which the unit rate always comes down to 3 in/sec. I write on the board and ask the students to copy in their notebooks:

For linear equations, equal changes in x produce equal changes in y.

"In other words, for linear equations, there is a constant rate of change. Now let's graph each one on the same set of axes so we can see what this means." (The lighter grey line in Figure 23-1 represents 3 in/sec change; the darker grey line represents 2 in/sec change.)

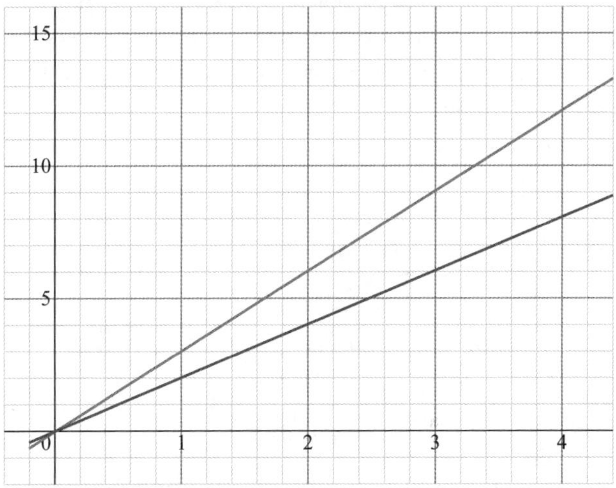

Figure 23-1

"Which line would you say is steeper?" There is a general consensus that the 3 in/sec line is steeper.

"What these lines represent is a change in rate. The 3 in/sec line covers more distance for the same amount of time. We call that rate 'slope' of the line."

Possible Confusion

Observant students might ask, "Isn't the change always going to be the y value divided by the x value?"

The answer is that it will for proportional equations, which we discussed in a previous lesson. (See Part II, Chapter 14.) I explain that we will see what happens for non-proportional equations in a few minutes.

SLOPE

"As we've seen, slope tells us how much the vertical distance, which is called the 'rise,' changes compared to the horizontal distance, called the 'run.' Slope is sometimes called 'rise over run' or vertical distance over horizontal distance."

I describe another situation: A bottle contains 15 gallons of water and leaks at the rate of 5 gallons per minute. I show students a graph that illustrates this (see Figure 23-2). The x-axis represents minutes, and the y-axis represents gallons.

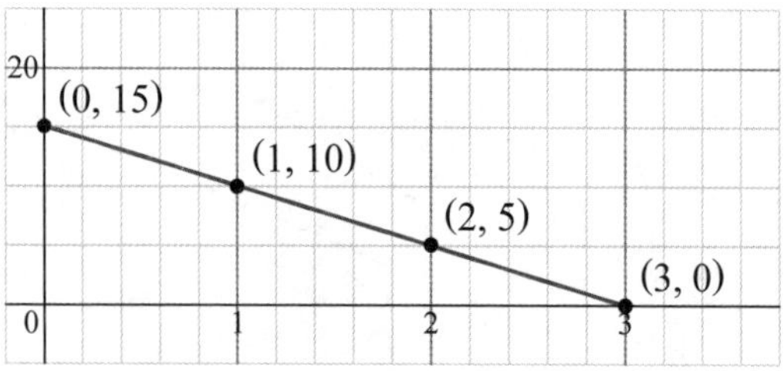

Figure 23-2

"The equation for this graph is $y = 15 - 5x$. We will find the slope of this line using a procedure that applies to any graph of a linear equation." I write the steps on the board and proceed to do a worked example as students write in their notebooks:

Step 1: Choose any two points on the line.

I'll pick several students so that we have two or three pairs of points, say (0,15) and (2,5), (1,10) and (3,0), and (0,15) and (3,0).

Step 2: Find the rate of change between the points.

"Let's pick two points: (0,15) and (2,5). The y-axis represents gallons, and the x axis represents minutes. So, the y coordinate of each ordered pair represents gallons. Let's find the change." I call on someone to tell me what the change is. I hear 15 – 5 = 10 and move on, doing the same for the change in time. "We subtract in the same order," I advise. The x coordinates are 0 and 2. "What did we get?"

There may be some hesitance to say −2.

"It's OK that it's negative. The change in gallons per change in time is $\frac{10}{-2}$. What is 10 divided by −2?"

Hearing –5, I pronounce that the slope is –5 gal/min. "What does the negative mean in this case? What's happening to the amount of water in the bottle?"

I hope to hear "It's decreasing" or "It's losing" and other equivalents. If I don't hear this, I offer prompts: "How do I represent a gain of five gallons/minute? Then, what does –5 gal/min mean?"

"That decrease is represented by a slope that is negative. The change in the amount of water in the bottle decreases by the same amount over the same unit of time."

This now provides the entrée for introducing the formula for finding the slope of any line.

"We just saw that we get the same slope for any two points that we pick on the line. We computed the change in y, which gives us the rise. And we divided by the change in x, which gives us the run."

I write the following on the board: *Slope* $= \frac{\text{rise}}{\text{run}} = \frac{\text{change in } y}{\text{change in } x}$.

"We found the change in y and the change in *x*, by subtracting the *y* and *x* coordinates:

$$m = \frac{\textit{difference of y coordinates}}{\textit{difference of x coordinates}} = \frac{y_2 - y_1}{x_2 - x_1}$$

"We use the letter *m* to represent slope. And we use the little numbers, called 'sub-scripts,' to refer to the two *y* and two *x* coordinates. Let's look at the example again" (see Figure 23-3).

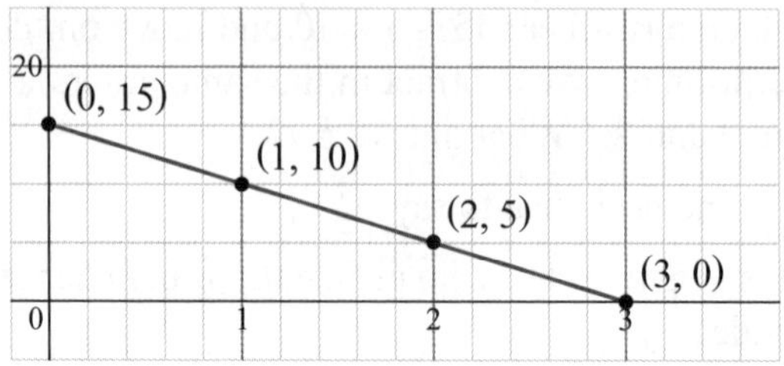

Figure 23-3

"Let's look at (1,10) and (2,5). To keep track of the coordinates, we can do this:

$$(x_1, y_1) = (1, 10); (x_2, y_2) = (2, 5)$$

"What is the value of x_1?" I should hear, "one," and I do the same thing for all the coordinates until students are familiar with how the naming scheme works. "Now, let's plug these into the formula." We get: $\frac{5-10}{2-1} = \frac{-5}{1} = -5$.

I hear some objections that the order of the numbers was different than how I did the first example.

"You're right!" I say. It doesn't matter what order you use, as long as you keep the same order in the numerator and denominator. Let's do it the opposite direction: $\frac{10-5}{1-2} = \frac{5}{-1} = -5$.

Examples. We put it to work with examples. I point out that we don't need to see the graph in order to find the slope. We just need to apply the formula.

1. Find the slope of the line that passes through (−4, 3) and (1,2).

   Answer: $\frac{1}{-5}$ or $\frac{-1}{5}$. *I make clear that either one means the quotient or resulting fraction is negative*; $-\frac{1}{5}$ is how we write it.
2. (1,-5),(8,3). *Answer*: $\frac{-8}{-7}$ or $\frac{8}{7} = \frac{8}{7}$.
3. (1,1),(3,5). *Answer*: $\frac{4}{2}$ or $\frac{-4}{-2} = 2$.
4. (3,4) *and* (2,4). *Answer*: $\frac{0}{-1} = 0$. *Students think they have done something wrong. I explain that a slope of 0 means a horizontal line, and we graph it to show that this is true.*

5. (3,4) *and* (3,5). *Answer:* $\frac{1}{0}$ *, which is undefined. We graph it to show it is a vertical line. "Vertical lines have a slope that is undefined."*

Horizontal and vertical lines are covered in a subsequent lesson.

Homework

Homework problems should consist of finding the slope of lines presented on graphs, in tables, or as a pair of points, as shown in the examples. For lines on graphs, the points should be marked as we did in the examples, rather than having students find where the line intersects with grid points, as is sometimes done in textbooks.

24. SLOPE-INTERCEPT FORM OF LINEAR EQUATIONS, AND GRAPHING

In a prior lesson, students learned that horizontal lines have a slope of zero, and vertical lines have a slope that is undefined. They have also learned how to find the slope of a line by inspection; that is, counting the vertical and horizontal distances between two points to find the "rise over run."

In the interest of full disclosure, I confess that I've always been surprised at how students seem to have difficulty remembering what's what in the equation: $y = mx + b$. What does work is practicing identifying what is the y-intercept (b) and slope (m) of the line described by the equation—as well as graphing the equation using this information.

Typically, textbooks ask students to transform equations in standard form ($Ax + By = C$) into the slope-intercept form when graphing the equation. This has its usefulness, but is somewhat limited at the seventh-grade level since students have difficulty solving for y for an equation like $3x + 2y = 7$. Solving it requires dividing all terms by 2, which seventh-graders find confusing. Textbooks generally limit such exercises to equations where y has no coefficient. I offer the more involved equation as a challenge for those students who can do it. Eighth-graders taking Math 8 do better at this. In algebra classes, I require all students to be proficient in this.

WARM-UPS

1. Express $\frac{1}{40}$ as a percentage. *Answer*: $1 \div 40 = 0.025 = 2.5\%$.

2. Find the slope of the graph. *Answer*: $\frac{-3-(-2)}{2-4} = \frac{-1}{-2} = \frac{1}{2}$. *Slope can also be obtained by inspection; vertical distance or rise = 1, and run = 2.*

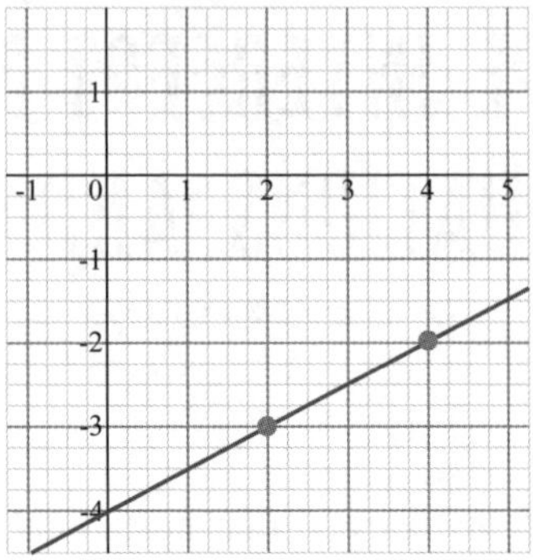

3. What is the y-intercept in the graph above? *Answer*: −4.
4. A purse was discounted 15% and sold for $51. Find the original price. *Answer*: $\frac{100-15}{100} = \frac{51}{x}$; $\frac{0.85}{1} = \frac{51}{x}$; $0.85x = 51$; $x = \$60$.
5. Find the x and y intercepts for the equation $3x - 2y = 9$. *Answer*: x intercept: $3x = 9$, $x = 3$; y intercept: $-2y = 9$, $y = -\frac{9}{2}$.

Problem 2 requires students to find the slope of the line; two of the points are indicated but unlabeled. Students will need to identify the coordinates of the points in order to apply the scope formula, or they can count the vertical and horizontal units by inspection. Problem 3 calls for identification of the y-intercept, which will be a part of the day's lesson to follow. Problem 4 is a review of finding the original amount of an item when the discounted price and discount rates are known.

THE SLOPE AND Y-INTERCEPT

"As you've seen, there are a number of ways you've learned to graph an equation. Today we'll learn yet another one."

I wait for groans and then proceed. "It is probably the one you will use most often."

Dramatic pause.

"Because it's the easiest," I say, and continue. "Let's start out graphing an equation from a table of values." The students graph it from this table (see Figure 24-1).

x	−1	0	1
y	1	4	7

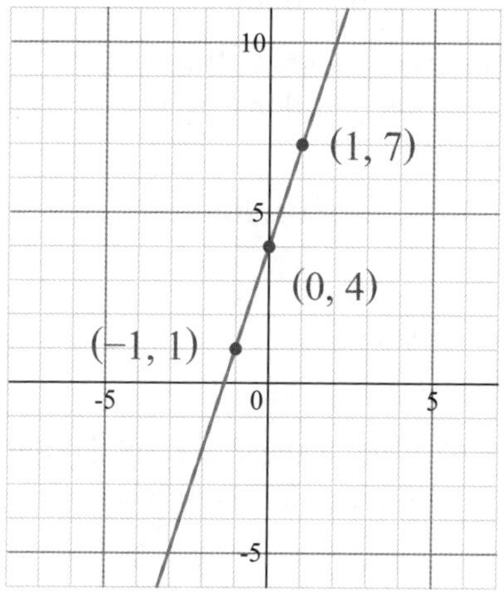

Figure 24-1

"What is the slope of the line?"

Response: The slope is 3.

"Now someone tell me what the y-intercept is."

Response: The answer is 4.

"I happen to know what the equation of this line is. It's $y = 3x + 4$. We said the slope of the line is 3. What is the coefficient of x in the equation?"

They see that it is 3.

"Now what do you think the 4 in the equation represents?"

They identify 4 as the y intercept, which is my cue to move on.

THE SLOPE-INTERCEPT EQUATION

"In the equation $y = 3x + 4$, we saw that the coefficient of x in the equation is the slope of the line, and the constant in the equation, 4, is the y intercept of the line. This is not a coincidence. The equation $y = 3x + 4$ is in what is called slope-intercept form. All straight lines can be represented by equations in the following form: $y = mx + b$.

"The m in the equation is the slope of the line, and b is the y-intercept. If an equation is in that form, you don't even have to plot it to know what the slope and y intercept are.

"Somebody tell me the slope of the line $y = -2x + 9$."

Response: −2.

"That leaves 9 to be the what?"

Response: The y intercept.

"What about $y = 3x - 9$? What is the slope and the y-intercept?" The slope they get right away, and I pause for the y intercept.

"Suppose I were to write it this way?" $y = 3x + (-9)$.

"Remember, subtraction is addition of the opposite." I usually see more students raising their hands to volunteer the answer: −9.

EXAMPLES

Find the slope and y-intercept of the equations.

1. $y = 5x + 3$. *Answer*: $m = 5$, $b = 3$.
2. $y = -6x - 8$. *Answer*: $m = -6$, $b = -8$.
3. $2x + y = 5$. This will give them pause. Prompt: "We need to put this in $y = mx + b$ form. How can we do that?" Someone usually volunteers that you can solve for y, which we then do, and I advise them to put the x term first on the right-hand side so it is more obvious what is m and what is b. They get $y = -2x + 5$; $m = -2$, $b = 5$.
4. $3x = y + 9$ *Answer*: $y = 3x - 9$; $m = 3$, $b = -9$.

Common mistakes: Students will confuse the m and b values. Some students may still be weak at solving for y.

Graphing Equations in Slope-Intercept Form

Having worked through examples of the slope-intercept equation to identify slope and y-intercept, I turn to graphing equations in this form.

I start with an equation and ask students to work in their notebooks as I work on the board. $y = 2x - 1$.

I then write the steps as we work through the first one.

Step 1: Identify slope (m) and y intercept (b): $m = 2$, $b = -1$.

Step 2: Plot y intercept. (0,-1).

See Figure 24-2.

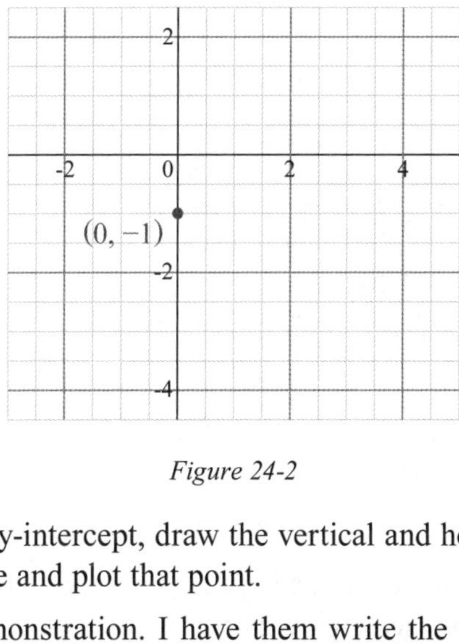

Figure 24-2

Step 3: From the y-intercept, draw the vertical and horizontal lines that represent the slope and plot that point.

This requires demonstration. I have them write the slope as $\frac{2}{1}$ so they know that we are going to draw a vertical line 2 units up from the intercept and then 1 unit to the right. "For drawing verticals, positive numbers mean 'up,' and negative numbers mean 'down.' When we draw horizontals, positive numbers mean go right, and negative numbers mean go left" (see Figures 24-3 and 24-4).

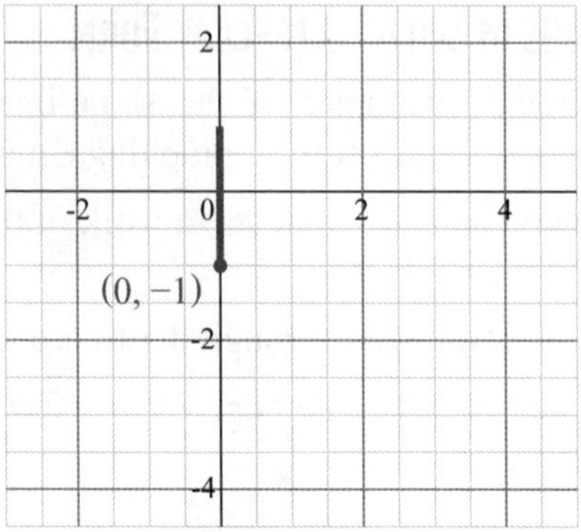

Figure 24-3

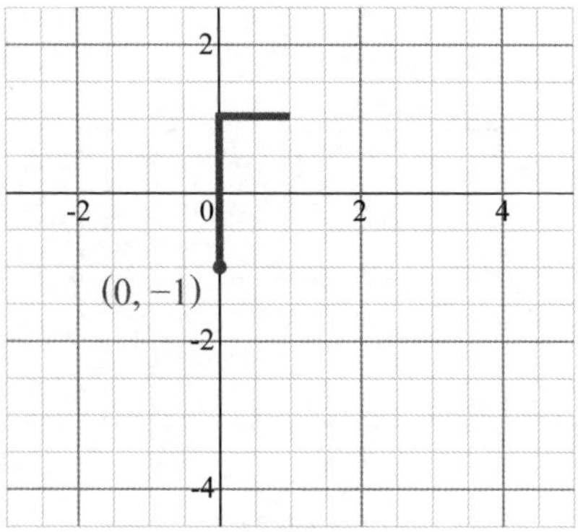

Figure 24-4

Step 4: From the point obtained in Step 3, use the slope in the same way to find another point (see Figure 24-5).

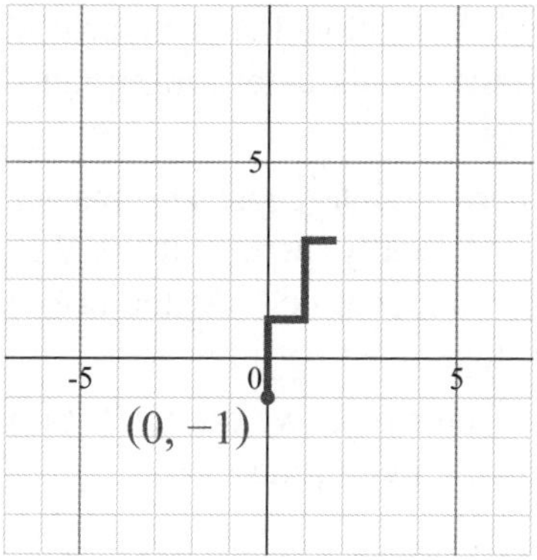

Figure 24-5

Step 5: Connect the three points with a straight line (see Figure 24-6).

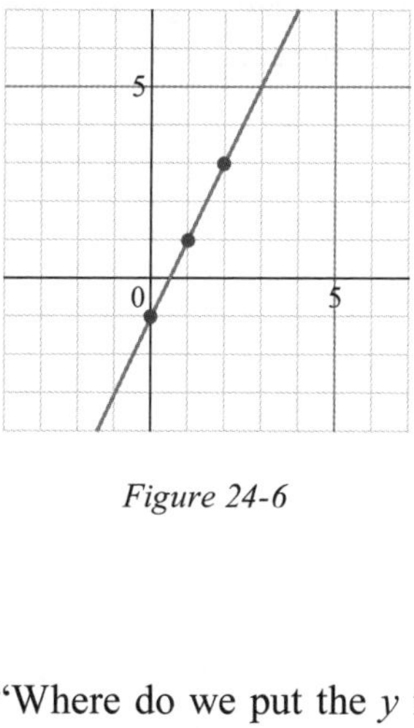

Figure 24-6

Examples.

1. Graph $y = -3x + 2$ "Where do we put the y intercept?" It is plotted at (0,2).

The slope (m) is −3, which I instruct students to write as $\frac{-3}{1}$.

"We said positive numbers are up, and negative numbers are down. So, from (0,2), we go 3 units in what direction?" Down.

"Now, we go 1 one unit to the right or left?" Right.

"Repeat the process." They do so and connect the points.

2. Graph $y = -\frac{2}{3}x + 2$. We follow the same procedure by plotting the y intercept, plotting the points determining the slope of $-\frac{2}{3}$, and then repeating the process.
3. Graph $y = 4x$. Is there a y-intercept? The answer I usually get is "no," to which I respond "Actually there is. It's 0. We just don't write the 0. So, where does the intercept go?" At the origin. Then, students plot the points for the slope.
4. Graph $3x + y = 5$. "What must we do first?" Solve for y. The equation is $y = -3x + 5$, which they will do.

Homework

The homework is more of the same, and as usual, I leave time to provide guidance and help as needed. In the process, I might pick a problem to do together that has a slope of, say, $-\frac{3}{4}$. "We can write this as $\frac{-3}{4}$, but we can also write it as $\frac{3}{-4}$." We proceed with that so that they see points can be plotted either way.

25. SOLVING SYSTEMS OF EQUATIONS BY GRAPHING

In this lesson, the components of linear equations that have been presented previously come together in solving systems of linear equations. All the points on the graph of a linear equation represent all the solutions of that equation. This point, although stated on various occasions in this unit, becomes more obvious (or should become so) when we start to graph a system of two equations. The point where the lines intersect represents the x and y values common to the two equations.

Graphing the equations makes this point clear. I must confess, however, that I dislike this lesson because graphing to find the solution to a system of equations is time consuming and frequently imprecise—which students very quickly discover. This discovery then serves as an entrée to subsequent lessons that provide instruction for more efficient techniques. I try not to prolong the agony, and I assign very few problems, but enough so that some of the problems lend themselves to neat and easy solutions, whereas others do not. Students get the point quick enough.

As part of the lesson, I also focus on systems that have no solutions (represented by parallel lines) and infinite solutions (represented by a single line).

Warm-Ups

1. Solve for y. $2x + 4y = 8$. (*Hint*: Divide *all* terms by the coefficient of y.) *Answer*: $4y = -2x + 8$; $y = -\frac{1}{2}x + 2$.

2. Identify the slope and y-intercept of the equation. (*Hint*: Divide *all* terms by the coefficient of *y*.) $2x + 3y = -9$.

   Answer: *Solve for y*: $3y = -2x - 9$; $y = -\frac{2}{3}(x) - 3$; $m = -\frac{2}{3}$, $b = -3$.

3. Graph the equation. $-2x + y = -5$.

   Answer: *Solve for y*: $y = 2x - 5$; $m = 2$, $b = -5$.

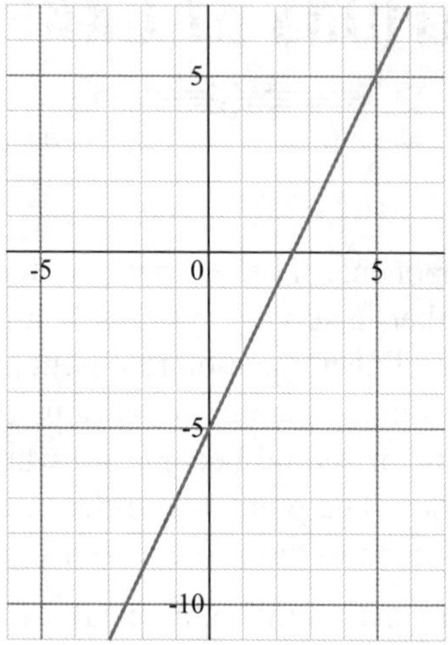

4. Simplify. $\frac{2}{3}(9x + 3)$. *Answer*: $\frac{2}{3}\,9x + \frac{2}{3}(3) = 6x + 2$.

5. Find *y* if $x = 1$ for $y = -3x + 4$. *Answer*: $y = -3(1) + 4 = 1$.

Students are used to solving equations for *y* when it has a coefficient of 1. For Problems 1 and 2, *y* has a coefficient other than 1. For Problem 1, after isolating the y term, we then have $4y = -2x + 8$. Each term must then be divided by 4. For the *x* term, this means it will be $-\frac{2}{4}x$, which reduces to $\frac{1}{2}x$. For Problem 2, it is similar. I walk them through Problem 1 and then ask them to work independently on Problem 2, but I will provide guidance as necessary.

The Graph of a Linear Equation Represents its Solutions

Problem 5 of the Warm-Ups has students plugging in 1 for *x* to obtain the value for *y*. Referring to this problem, I point out that we have solutions for x and y: 1 and 1. "If we were to write this as an ordered pair, that is (x,y), how would we write it?"

If faced with blank stares, I refrain from saying, "It's like asking who's buried in Grant's Tomb?" mainly because despite the simplicity of the question, students may be overwhelmed, and secondly, the response "Who's Grant?" that I sometimes get is just as disturbing as their not seeing what the answer is. They quickly catch on that the question is fairly simple and say, "(1,1)."

"Now suppose I ask how you can tell if the ordered pair (2,1) is on the line for $y = -3x + 4$?"

If there is no response, I'll prompt with "What does the ordered pair tell us about what y equals?" and "What does y equal when x equals 2?"

The answer is –2, which clearly does not equal 1. "So, is (2, 1) on the line?" The general consensus is "no."

"Right. To be on the line, y must equal – 2 when x equals 2."

"Now, let's say we have the equation $x + y = 2$; this is equivalent to $y = -x + 2$." I ask students to graph the line.

"The graph of this line is made up of infinitely many points. There are many solutions to this equation; that is, there are many ordered pairs that make the equation true. Is the point (1,1) on this line? Does it make the equation true? Work it out." In fact, (1,1) is on the line since $1 = -1 + 2$.

"Plot the point (1,1) so you can see where it is on the line" (see Figure 25-1).

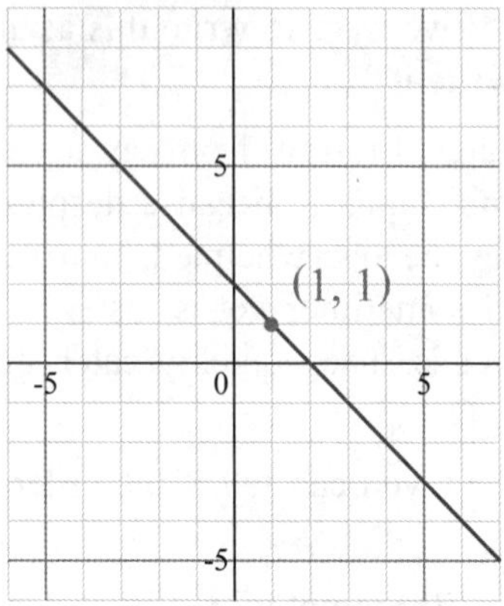

Figure 25-1

"Now, graph the equation $x - y = 0$. Is (1,1) on that line?" (To do this, they solve for y to obtain $y = x$.)

"We can see that the point (1,1) is on both lines. Just to make sure, plug (1, 1) into $x - y = 0$ to see if it checks out" (see Figure 25-2).

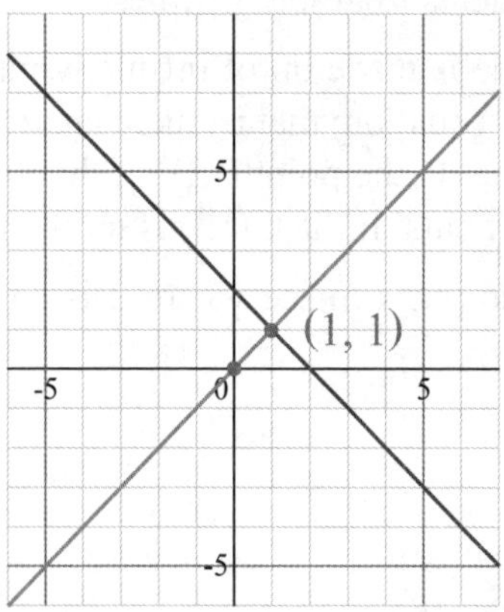

Figure 25-2

SYSTEMS OF EQUATIONS

After graphing the equations, I state that the point of intersection is the ordered pair that is on each of the lines. "It is the only point that is in common with the two lines. This is a system of equations," I say, and I write on the board:

System of Equations: A collection of two or more equations with the same set of variables.

"We can use graphing to find out the common solution for these equations. That is, we can find values for x and y that make these equations true." In the wake of the silence that inevitably follows this statement, I add: "In other words, we can use graphing to solve for x and y."

Examples. I ask students to find solutions to systems by graphing in their notebooks. The first one we do together, and the others they work on independently while I walk around to provide guidance. I show student work with a document camera.

1. $x + y = 4$; $2x + y = 5$. *Answer*: (1,3).
2. $x + y = -1$; $y = 3x + 7$. *Answer*: (–2,1).
3. $y = -3x + 2$; $y + 3 = 2x$. *Answer*: (1,–1).

For fractional answers, I tell students that their answer does not have to be exact; some estimation is required. I work with them to help them estimate where the points are.

4. $y = -x - 1$; $y - 4 = x$. *Answer*: $(-2\frac{1}{2}, 1\frac{1}{2})$.
5. y =1/4 x -3; y =-1/4 x – 2.Answer: $(2, -2\frac{1}{2})$.

No Solution, Many Solutions

I ask students to graph the following equations on the same set of axes:

$$y = x + 1; y = x - 1$$

"Do the graphs intersect?" In fact, they do not. They are parallel lines.

"Sometimes a system of equations does not have a solution; if the equations have the same slope, which these do, they are parallel. Parallel lines do not intersect. So, we say there is no solution."

Next, I write two equations on the board: $y - 4 = 2x$ and $y - 2x = 4$. Some students, seeing that the slopes are the same, will say "No solution; they're parallel." I reply, "Take a closer look."

Soon after, someone else will say, "They're the same equation." When the equations are the same, then there is only one line. As discussed at the beginning of the lesson, a line represents all the possible solutions to the equation the line represents. "We say that there are infinitely many solutions."

Homework. Problems will be similar to what we did in class. I limit the number of problems to no more than 10.

26. SOLVING SYSTEMS OF EQUATIONS BY ELIMINATION

When going over the homework for the previous assignment, students undoubtedly ask about problems that have fractions as solutions or have fractions in the equation. It becomes obvious that graphing is not the most efficient method for solving systems of equations. Also, even for problems for which the solutions are integers, the construction of the lines in the graph have to be done carefully.

In one class, I asked students what they thought of the method of graphing to solve equations. One boy said he liked it, and I asked why. "You can see what's going on all at once," he said. I found the answer insightful and remarked that indeed graphing does give you the big picture of what's going on, even if it is not the best way to solve equations.

ELIMINATION METHOD

This lesson focuses on solving systems of equations by the elimination method. Some textbooks introduce the substitution method first. Students find substitution confusing, and I've found that teaching the elimination method first is a better introduction.

I split up the elimination method into two lessons. The first is for problems in which the elimination does not require transformation by multiplication of any of the equations. That is, there are variables in both equations that can be cancelled by adding or subtracting the equations. For other equations in which there aren't such variables (e.g., $3x + 2y = 10$; $5x - 3y = 4$), one or both equations must be multiplied by an integer to allow for the elimination step to be carried out. I also re-introduce word problems that previously were solved with one variable but asked to find two unknowns. This time, I show how they can be solved using two variables. (I have not included in this chapter the elimination with multiplication method; the word problem lessons are presented in the section on eighth-grade algebra.)

SUBSTITUTION METHOD

Solving systems by substitution is generally a confusing topic for seventh-graders as well as eighth-graders. (An example lesson for the substitution method is not included in this book.) The approach I've used is to show how some systems lend themselves to substitution more than elimination. Specifically, this pair of equations lends itself to substitution: $y = 4$; $x - y = 5$.

Students can see that four can be substituted for the y variable in the second equation. Similarly, this pair also lends itself to substitution:

$$y = x + 4$$

$$x + y = 16$$

The first equation establishes that $x + 4$ can be used in place of y in the second equation.

Systems of equations that lend themselves to elimination but which students are required to solve by substitution can be confusing. For example:

$$x + 2y = 8$$

$$x - y = -1$$

To solve by substitution, one possibility is for the first equation to be solved for x: $x = 8 - 2y$. The expression $8 - 2y$ is then substituted for x in the second equation: $8 - 2y - y = -1$. Students have difficulty with this at first. Ultimately, students use the method of their choosing based on what appears to be the most straightforward approach.

Warm-Ups

1. A certain number plus twice that number is –24. Find the two numbers. *Answer: Let* x *= a number;* $2x$ *= twice that number.* $x + 2x = -24$;

   $3x = -24$, $x = -8$, $2x = -16$.
2. Find the slope of the line that passes through (-3,4) and (-5,4). *Answer: 0.*
3. $-5 + 5 = ?$ *Answer: 0.*
4. Without graphing, how many solutions does this system of equations have?

   $y = 2x - 5$, $y - 2x = 6$.

   Answer: None; the slopes are the same,so the lines are parallel.
5. Bob and Alicia's weights average 57 pounds. If Alicia weighs 50 pounds, how much does Bob weigh? *Answer: Total weight* = 57 × 2 = 114 *lb.*

   Bob's weight: 114 – 50 = 64 *lb.*

Problem 1 will be used later in this lesson to show how such problems can be solved with two variables. Problem 3 is a reminder that adding opposites equals 0, which is what happens in the elimination method.

Elimination of a Variable by Addition

Students by now have learned that when solving equations, we sometimes use the additive property of equality.

"Suppose I have the equations $x + y = 5$ and $x - y = 2$. I want to solve this but not by graphing. We all agree that we can add the same amount to both sides of an equation, do we not?" Agreement ensues.

"So, if I wanted to add 10 to both sides of the equation, I could. It would look like:

$$x + y = 5$$

$$10 = 10$$

$$x + y + 10 = 5 + 10$$

"I don't really want to add 10 to both sides; I'm just making a point. Since we can add equal amounts to both sides of an equation, there is something I'd like to add to both that will help me solve it. And what I'd like to add is the other equation: $x - y = 2$. This equation says that the left side of the equation, $x - y$, is the same number as 2, does it not? Isn't that what the equal sign means?"

They agree, but they wonder where I'm going with this.

"Since I have an equal amount on both sides, I can add this to the first equation:

$$x + y = 10$$
$$+(x - y = 2)$$

"Can anyone tell me why I would want to do that?"

Response: y and $-y$ will cancel.

"Exactly. The y's disappear because addition of opposites equals zero. So, let's add the equations. What's $x + x$? What's $10 + 2$?"

We end up with $2x = 12$, which I write on the board. Students then get the answer quickly: $x = 6$. "Correct. Now that we know what x is, how can I use that to solve for y?"

I wait for someone to tell me I can plug it into an equation. If no one answers, I continue.

"Can I substitute 6 for x in the first equation? Let's do it." I write $6 + y = 10$. Students solve it. $y = 4$.

"Let's see if that works in the second equation: $6 - 4 = 2$. Yes, it checks. That's the answer."

Once in a while, a student will say that the equation was so easy they were able to figure it out without going through all that; that is, they used "guess and check."

"Yes, but it isn't very efficient. Suppose we had these equations."

$$2x + y = 200$$
$$x - y = 40$$

"Would you want to use guess and check? I wouldn't. Let's solve it. Tell me the steps."

I call on a student, hand up or not. I'm told to add. "Why?"

Response: To cancel the *y*s.

We obtain $3x = 240$; $x = 80$. "Let's plug this into the top equation."

$$160 + y = 200; y = 40$$

Subtracting Equations

I write the following system on the board: $x + 4y = 27$; $x + 2y = 21$.

Students will look at the *x*s and say we should subtract.

"Yes, you can subtract, and that would be correct. But remember: Subtraction is the same thing as adding the opposite. I find students make fewer mistakes when adding than subtracting, so I want to keep everything in addition mode. Let's multiply one of the equations by –1."

I show how we do this, picking the top equation, though I can pick either one. I do it one step at a time, showing what we get when we multiply each term by −1. We end up with $-x - 4y = -27$. "Now, add the two equations."

I circulate around to make sure students are doing this correctly, offering help when needed. The sum is $-2y = -6$.

Dividing both sides by –2, not 2 (a common mistake), students get $y = 3$. "Now substitute *y* in one of the original equations." Checking their solutions as I walk around the room, I see they get $x = 15$.

Examples. I work out the first few problems with the class so they are comfortable with the procedure. I then select students to work out problems on the board.

1. $x + y = 74$, $x - y = 16$. *Answer*: $x = 45$, $y = 29$.
2. $x + y = 122$, $x - y = 28$. *Answer*: $x = 75$, $y = 47$.
3. $x + 6y = 10$; $x + 2y = 2$.

   Answer: $x + 6y = 10$; $-x - 2y = -2$; $4y = 8$; $y = 2$, $x = -2$.

For this problem, I remind students that one of the equations should be multiplied by –1.

4. $3p - q = 10; 2p - q = 7.$

   Answer: $3p - q = 10; -2p + q = -7; p = 3, q = -1.$

HOMEWORK

Problems are more of the same. As necessary, I may work one or two of the homework problems as extensions of examples to ensure that students understand the procedure.

PART III.
EIGHTH-GRADE ALGEBRA

Algebra in eighth grade is generally reserved for students who qualify by having taken accelerated Math 7 and/or qualifying via a placement exam. There are some students who may be placed in algebra who did not take the accelerated Math 7 course, but did well on the placement exam and otherwise show promise. In addition, there are students who are placed in algebra in eighth grade at the insistence of parents, but could benefit from a year of Math 8.

I have written elsewhere that I use "Modern Algebra", by Dolciani et al. (1962), copies of which I obtained from the Internet when the price was a lot less than its current one ($88 as of this writing). I like the book for its sequencing of topics and for the problems it presents. Some of the explanations in the book are very formal—an artifact of the 1960's New Math sensibility that prevailed when the book was written. The example lessons included in this book provide the dialogue I add around the nucleus of the algebra textbook.

For those students who might be lost for the reasons indicated above, I have often had them use a more algebra-lite type of book, also called *Basic Algebra*, by Brown, Smith and Dolciani (1988).

Students have had some experience solving equations in seventh grade; accelerated math students will have had more. In any case, I handle basic equation solving as a review and spend more time on equations with variables on both sides, those with fractional coefficients, and those that require distribution within the equation itself. Students will have learned how to operate with negative numbers, so that's another topic that does not need to be repeated.

Many of today's algebra textbooks have a dearth of word problems, and those that are included are usually of the "real-world" variety, which tend to be wordy and often do not generalize to other problems having similar structure. I focus quite a bit on word problems. Because students have difficulty with these, my tests do not include a lot of them—but there will be two or three. Those which I do include are relatively straightforward. I provide more complex word problems for extra credit so that there is a differentiation of problem types on tests. This practice is not differentiated instruction; I provide the same instruction to all students. Students with higher ability than others tend to do the extra credit problems. These problems only add to the score on a test or quiz. They carry no penalty; all students are welcome to try them.

In my own experience, my algebra teacher recognized the difficulty students have with word problems. He admitted that he was not very good at solving or teaching them. When I took algebra 2, however, that teacher provided thorough instruction, and I was able to follow. This might have been because of better instruction, but I believe that it was also an artifact of the maturity that comes with consolidation of algebraic procedures and concepts taught in algebra 1 that have had time to sink in.

In a first course in algebra, students are not likely to make the connections they are able to make later in solving word problems. For example, there are many types of distance/rate problems, some of which can be solved by a "distance = distance" approach, and others with a "time = time" approach. And even those solved using "distance = distance" have variation; that is, travel in opposite directions, travel in the same direction, and round-trip travel. I provide a mixture of such problems in Warm-Ups, homework, and quizzes, but I try to keep it simple. The end result has been that my students generally can solve more word problems than others who are not subjected to the variety of traditional math problems.

Following below is the scope and sequence of an eighth-grade algebra course. Most of the material in the first two units have been covered in accelerated Math 7, and Math 8. Bolded topics in the list below indicate the example lessons included in this book. I do not spend a great amount of time on the italicized topics (Graphing Quadratics, and Exponential Functions). They are presented again in algebra 2 and pre-calculus. Because state exams that are aligned with Common Core do not address algebra until the eleventh grade, there is plenty of time for students to become proficient in these areas.

Typically, I reach these units at the end of the school year when attention is waning and students are thinking "I can't take in any more new information." I strive to reach the unit on quadratic equations by April, which is before the "I can't take any more" syndrome sets in. Quadratic equations are, in my opinion, the pinnacle of an algebra 1 course, particularly the derivation of the quadratic formula.

Scope and Sequence of Topics for Eighth-Grade Algebra

The numbers indicate the chapter number for the example lessons in this part. Lessons contain a brief summary of the content of topics that occur prior to the particular lesson and which pertain to the particular lesson, but are omitted from the book.

Variables and Open Sentences

Identifying Factors, Coefficients and Exponents

Combining Like Terms: Addition and Subtraction

Solving Open Sentences (Equations)

Thinking with Variables: From Words to Symbols

1. Solving Word Problems

2. Multiplication (Exponents; Powers)

Division (Exponents; Powers)

Equations

Basic Axioms

Distributive Property; Special Properties of 1 and 0

Review of One- and Two-Step Equations

Solving Equations with Variables on Both Sides

Solving Equations, Inequalities and Word Problems

3. Literal Equations

Absolute Value Equations

Solving Inequalities

4. Solving Absolute Value Inequalities

Problems about Consecutive Integers

5. Uniform Motion Problems: Opposite Directions

6. Uniform Motion Problems: Same Direction

Polynomials

Adding Polynomials

Subtracting Polynomials

Products of Powers

7. Power of Products and of Powers

Multiplying a Polynomial by a Monomial

8. Multiplying Two Polynomials

Powers of Polynomials

9. Zero Power and Negative Exponents

Factoring and Special Products

Factoring in Algebra

10. Identifying Common Factors

Elimination Method with Multiplication

20. Word Problems

Substitution Method

Linear Inequalities

Digit Problems

21. Wind and Current Problems

Real Numbers

Properties of Square Roots

Repeating Decimals

Pythagorean Theorem

Rationalizing Denominators

Adding and Subtracting Radicals

Fractional Exponents

Solving Radical Equations

Quadratic Equations

22. Square Root Method of Quadratic Equations

23. Solving Equations by Completing the Square

24. Quadratic Formula

25. Derivation of Quadratic Formula

Discriminant

Functions and Variation

Definition of Function

Function Notation

Direct Variation

Inverse Variation

Joint Variation

GRAPHING QUADRATICS

Graphing Quadratics

Graphing y=ax^2+c

Graphing y = ax^2 + bx + c

Graphing Maximums and Minimums of Quadratic Functions

Vertex Form of Quadratics

EXPONENTIAL FUNCTIONS

Review of Square Root Properties

Review of Quotient of Powers; Negative Exponents

Exponential Functions Part I

Exponential Functions Part II

Exponential Growth

Exponential Decay

Geometric Sequences

1. SOLVING WORD PROBLEMS

Prior to these lesson, students have had a review of basic equation solving which they have had in seventh grade. They also have had some more complex multi-step equations involving distributions and variables and numbers on both sides of the equation.

Students have also had instruction with some simple word problems. A typical problem is: "The length of a rectangle is five times its width. The perimeter of the rectangle is 240 ft. Find the length and width of the rectangle."

Students' difficulties with word problems lie in finding out how one interprets and assembles information in the problem as an equation. This requires knowing what is being asked: that is, what unknowns are to be identified, and what quantities are being equated.

When I taught algebra, I had a poster on the wall that read:

Teacher: Suppose x is the number of sheep.

Student: But what if x is not the number of sheep?

The poster was more of a reminder to me than to my students that part of the skill in solving problems is deciding how to represent quantities in terms of a variable. In a problem with two variables such as a smaller and a larger number, one can define x to be either one; the remaining unknown then is written in terms of how x has been defined.

In this lesson, the problems become a bit more involved and will continue so as the course progresses. Rectangle problems will eventually include the concept of area so that by the end of the course, students are having to solve problems such as: "Two tin squares together have an area of 325 square inches. One square is 5 inches longer than the other. Find the side of each."

I view word problems in this course as a way for concepts and methods for solving various problem types to remain as a muscle-memory of the mind to be re-energized in subsequent math courses. I recognize this will not always be the case for all students. There is no royal road that substitutes for hard work and practice.

Warm-Ups

1. Solve. $5z = -2z + 14{,}021$. *Answer*: $7z = 14{,}021$; $z = 2{,}003$.
2. Solve. $8x - 5 = -5$. *Answer*: $8x = 0$, $x = 0$.
3. Solve. $2(t + 4) - 3 = \frac{1}{2}(10 + 6t)$.

   Answer: $2t + 8 - 3 = 5 + 3t$; $2t + 5 = 5 + 3t$; $t = 0$.
4. Write an equation and solve. If 2 times a number is decreased by 16, the result is 4. Find the number. *Answer*: $2x - 16 = 4$; $2x = 20$; $x = 10$.
5. Combine like terms. $-2x - 5y - 5x - 6$. *Answer*: $-7x - 5y - 6$.

These problems provide review of various types of equations. Problem 3 requires multiplication by $\frac{1}{2}$. A prompt may be "What is $\frac{1}{2}$ of 10? What is $\frac{1}{2}$ of $6t$?"

Expressing One Unknown in Terms of the Other

Rather than announce that today's lesson is on word problems (resulting in groans and expressions of disdain), I simply project the following problem on the board:

"The sum of the length and width of a rectangle is 42 inches. Twice the length is 1 inch less than 3 times the width. Find the dimensions of the rectangle."

"Who remembers the formula for perimeter of a rectangle?" I ask, praying that someone will remember. Someone usually does. (I make a mental note of how many are retaining things like this (and who they are) as a baseline.)

I write the formula on the board: $2L + 2W = P$.

"Can I start substituting into this formula? What information do I have? What is the perimeter? Read the problem and tell me."

I call on someone regardless of hand up or not. If the person doesn't know, someone inevitably whispers the answer to the person. Although I try to avoid such situations, I use this to again find out who is on top of things and who will need more help.

I'm eventually told the perimeter is 42 inches. Now, the formula looks like $2L + 2W = 42$.

"What should we let x represent? Length or width? Which would be easier?"

The idea is to get them to find a dimension for x so that the other dimension can easily be defined in terms of it. The general consensus is width.

"Let's write down the variable to keep track."

I write x = *width of rectangle.*

"How does the problem define the length? I hear "1 inch less than 3 times the width."

"How would I write length then in terms of width? Where do I start?" If no immediate response, I prompt: "If the width were 10 inches what is 1 inch less than 3 times 10 inches? Do I multiply first or subtract?"

Someone will say that you multiply 10 times 3 and subtract 1. "Ah ha. So, $10 \cdot 3 - 1$. Instead of writing 10, let's write x since that's the width, and what do we get?"

They should get $3x - 1$.

"Now, write the equation in your notebooks and I'll be coming around. Just the equation, that's all you have to do for now. The easy part is solving it. I want you to do the hard part now."

I'm looking for $2(3x - 1) + 2x = 42$. If students need help I offer prompts such as "What do I plug into L in the perimeter formula?"

Hiding in Plain Sight

I project the following problem on the board:

The sum of two numbers is 46. Two times the smaller number plus the larger number is 52. Find the numbers.

The resulting blank stares tell me they've read the problem.

"It might help to write this in English first. Look at the second sentence." I then write the following: *Two* · (*smaller number*) + *larger number* =

"What do I write on the right-hand side?" They say "52" and now the equation looks like:

$$\textit{Two} \cdot (\textit{smaller number}) + \textit{larger number} = 52$$

"Let's look at the first sentence now. It says the sum of larger and smaller is 46. I'll start: I'm going to let x equal the smaller number. Any ideas of how I represent the larger number?"

Some students may say "y," to which I respond: "I meant in terms of x." (Not that there's anything wrong with saying "y," but they are not quite ready to work with two variables—having tried that once. When we get to the chapter on linear systems, they are ready to take it on.)

"This is what I call a "hiding in plain sight" problem. Suppose the smaller number is 10; how would you find the larger number?" They say "Subtract 10 from 46."

I draw a diagram on the board:

10	46 − 10

"That gives me 10 and 36. Suppose the smaller number were five? What would I do?"

They tell me the answer and I update the diagram:

5	46 − 5

"You get the idea. Now, suppose the smaller number is x. How do I write the larger number?"

They see the pattern: $46 - x$.

"Let's plug things in. If x = the smaller number, then how do I write two times the smaller number?" I hear $2x$.

"Let's add the larger number."

We now have $2x + 46 - x = 52$. "Solve it." They do and get $x = 6$.

Example. I admonish students to write down how they are defining the variables.

1. Two numbers sum to 98. Four more than the larger number is twice the smaller number. What is the larger number?

I provide hints. "How do I write the smaller number?" What goes on each side of the equation?" *Answer: Let x = smaller number; larger number =* $98 - x.(98 - x) + 4 = 2x$; $3x = 102$; $x = 34$, $98 - x = 64$.

Some may question what would happen if we would have let x equal the larger number. In that case, the equation would be: $x + 4 = 2(98 - x)$; $x + 4 = 196 - 2x$; $3x = 192$; $x = 64$ (*larger number*) and $198 - 64 = 34$ (*smaller number*).

Doing this emphasizes the importance of writing down what the variables represent.

I continue with examples that I take from the homework to get them started on the problems in class. I provide help and guidance as needed.

Homework

Homework should be a mix of different types of word problems—among them, the "hiding in plain sight" type. I will work one or two of the more difficult homework problems with the class.

2. MULTIPLICATION (EXPONENTS, POWERS)

Like most topics in algebra, there are rules for the multiplication of monomials that contain powers; for example, $3x^4\ (4x^7)$. As familiar as these rules are to those of us who have been working with them for many years, students new to them may be overwhelmed.

When I was last teaching this topic, I brought up the difficulty with a fellow math teacher at my school who had been teaching algebra for 20 years. He told me that most algebra books present the rules of multiplication of powers in two short lessons, and then follow it up with the rules of quotient of powers, also in two short lessons. Even my beloved Dolciani et al. (1962) algebra textbook falls prey to this.

His solution was to spread out the information more slowly—teach the introductory material early and continue giving students problems involving exponent multiplication. Later, teach the more involved problems.

I therefore include two lessons at this early part of algebra. The subsequent lesson covers division rules for powers and builds on the logic established for multiplication. It is not included in this book.

Warm-Ups

1. Write an equation and solve. The length of a rectangle is 6 feet more than 3 times the width, and the perimeter is 188 feet. Find the dimensions of the rectangle.

   Answer: *Let* x = *width, then* $2(6 + 3x) + 2x = 188$; $12 + 6x + 2x = 188$; $12 + 8x = 188$; $8x = 176$; $x = 22$, and $6 + 3x = 72$.

2. The sum of two numbers is 78. If 3 times the smaller is increased by the larger, the result is 124. Find the smaller number.

   Answer: *Let* x = *smaller no.*; *then* $78 - x$ = *larger no.* $3x + 78 - x = 124$; $2x + 78 = 124$; $2x = 46$; $x = 23$.
3. Solve. $\frac{3}{10}t = 6$. *Answer*: $t = 6\left(\frac{10}{3}\right) = 20$.
4. Simplify. $-4(3x - 12) + 9x - 47$. *Answer*: $-12x + 48 + 9x - 47 = -3x + 1$.
5. Solve. $7(b + 2) - 4b = 2(b + 10)$.

   Answer: $7b + 14 - 4b = 2b + 20$; $3b + 14 = 2b + 20$; $b = 6$.

Problem 1 relies on students knowing the formula for the perimeter of a rectangle, for which some students will need reminding. Problem 2 is the "hiding in plain sight" type of problem talked about in the previous lesson. Even though it was talked about previously students still have difficulty recognizing when this structure is to be used. Problems 3– 5 are review of multi-step equations that students have had prior to this lesson.

EXPONENTS

Students have learned what exponents are in the seventh grade. I ask them what 8^3 represents. They usually remember, but will say it incorrectly. I often hear "It's 8 multiplied by itself 3 times."

Before I offer them the correct way to state it, I write on the board: $8^3 = 8 \times 8 \times 8$.

"The little number is called what?"

Response: An exponent.

"And the number 8 is called the base," I say. "It tells us that 8 is used as a factor 3 times. That is how you say it, by the way, not 8 multiplied by itself 3 times. How would you describe 7^5 in that same way?"

I want to hear "7 used as a factor 5 times," which I might hear mixed in with "$7 \times 7 \times 7 \times 7 \times 7$."

"This can also be called 7 to the power of 5, or 7 to the fifth power. Any number that has an exponent is called a power. How would I write a to the fourth power?" They write it on mini white-boards. I want to see a^4.

I write on the board: *aaaaaaa*… "This is *a* as a factor *n* times. Write this as a power." I want to see a^n.

"We all know that we write 3 times *a* as 3*a*. "What about $3 \cdot a \cdot a \cdot a$?"

I want to see $3a^3$.

Examples. I use the examples as tools of instruction.

1. $aa \times bb$. *Answer*: $a^2 b^2$.

Prompts may include: "How do I write *a* times *b*?"

2. $3aa \times bb$.Answer: $3a^2 b^2$.
3. $aaa \times bbbb \times cc$ *Answer*: $a^3 b^4 c^2$.
4. What does 2^5 equal? *Answer*: 32.
5. What does $\left(\frac{1}{2}\right)^3$ equal? *Answer*: $\frac{1}{8}$.
6. $2a \times 3a$.

This may take some prompts. This can be rewritten as $2 \cdot a \cdot 3 \cdot$ a which can be re-ordered as $2 \cdot 3 \cdot a \cdot a$. "What do we get when we multiply the numbers? And how do we write $a \cdot a$? Put it all together: $6a^2$

7. $2 \cdot a \cdot 3 \cdot a \cdot b \cdot 6 \cdot a \cdot b$. *Answer*: $36a^3 b^2$.

Common mistakes: Seeing $a \cdot 3$ and writing a^3.

Rule for Multiplying Powers

I write on the board: $a^2\ a^3\ a^4$.

"Let's write out what this means."

Starting with a^2 and continuing, I write: $aa \cdot aaa \cdot aaaa$.

"How many times is *a* used as a factor?" I'm told 9 times.

"How could I write this more compactly? That is, as *a* raised to what power?"

Students usually follow the logic and tell me a^9.

"What is $b^3 \cdot b^4$ then? *Answer*: b^7.

"So I think you can figure out what this is leading up to. When we multiply two powers that have the same base, instead of writing out all the factors, what's the quicker way of doing it?" There are usually more than a few students who will say "Add the exponents."

On the board and into their notebooks:

For all positive integers m and n: $a^m \cdot a^n = a^{m+n}$

"What about $a^2 b^3$? Can we simplify that?"

There will be some who say yes, and some who say no. I call on those who say no to find out why. Generally, they will say that the bases have to be the same since *aabbb* can't be combined into one base. I have them write in their notebooks:

Exponents are added only when the bases of the powers are the same.

Examples.

1. $a^2 \cdot a^5$. *Answer*: a^7.
2. $d \cdot d^2$. *Answer*: d^3.

A hint I give students is that the first d is the same as d^1; the 1 isn't written, just like we don't write $1x$. Some students will need to be reminded of this many times.

3. $x^2 \cdot x^3 \cdot y^2$. *Answer*: $x^5 y^2$.
4. *abab*. *Answer*: $a^2 b^2$.
5. $5xy \cdot 2x^2$. *Answer*: $5 \cdot 2 \cdot xx^2y = 10x^3y$.
6. $x^2 \cdot x^2 \cdot x^2$. *Answer*: x^8.

Homework

The homework should be fairly simple multiplication of powers like the examples. Some numeric problems should be included like:

$9^2 \cdot 9 \cdot 9^3$. Answer: 9^6; and $\left(\frac{1}{3}\right) \cdot \left(\frac{1}{3}\right) \cdot \left(\frac{1}{3}\right)$. Answer: $\left(\frac{1}{3^3}\right)$.

Problems should specify whether the answer is to be written in exponential form as above, or as a numerical value. For example, the numerical value of $9^3 = 243$.

3. LITERAL EQUATIONS

This lesson is the first part of a unit that introduces more complex equations, inequalities and uniform motion problems. The equations involve more negative numbers and fractions, both within the equation itself, as well as the solutions.

In this particular lesson the focus is on literal equations, for example, $I = prt$, in which components of the equation are letters. Understanding how such equations may be solved for other letters in the equation plays a role in seeing how one variable can be expressed in terms of the others. For example, the formula "distance equals rate times time" or $d = rt$, can be solved to express rate in terms of distance and time (i.e., $r = \frac{d}{t}$), as well as time in terms of distance and rate (i.e., $t = \frac{d}{r}$).

Throughout these lessons, I emphasize leaving improper fractions in fractional form and not converting to mixed numbers or decimals.

Warm-Ups

1. Write an equation and solve. Five times a certain number is 12 more than twice the number. What is the number? *Answer*: $5x = 12 + 2x$; $3x = 12$, $x = 4$.
2. Solve. $22 - x = 7x - 2$. *Answer*: $8x = 24$; $x = 3$.
3. An apple has twice as many calories as a peach. The two fruits together have 105 calories. How many calories are in each fruit?

   Answer: *Let* x = *the number of calories in a peach. Then* $x + 2x = 105$; $3x = 105$; $x = 35$ (*peach*) and $2x = 70$ (*apple*).
4. Solve. $2x + 5 = -101$ *Answer*: $2x = -106$; $x = -53$

5. Solve. $5y + 7 - 1 - y = 17$

   Answer: $4y + 6 = 17$; $4y = 11$; $y = 11/4$

Problem 4 requires students to subtract 5 from both sides. It may be important to remind them that subtraction of 5 is the same as –101+ (–5). Some tend to confuse it with −101 – (−5), and will add +5 to obtain –96. A helpful prompt to remind them is "I lost $101 and then lost $5; how much did I lose in total?"

NEGATIVE NUMBERS IN EQUATIONS

Problem 4 of the Warm-Ups provides an entrée to the subject of negative numbers in equations. "Anyone remember what the opposite of a number is? Like the opposite of 5 is what?" They should know that it is –5, but I also prepare myself for no one volunteering an answer.

"And what is the sum of a number and its opposite?" Zero. So far so good. Now, I up the ante.

"How about the opposite of x?" I'll usually hear "negative x," and asking for the opposite of $-x$, I'll hear "positive x." I'm always on the look-out for a student or students who may ask "But what if x is already negative?"

"Good question. Let's look at that. If x is −5, then $-x$ is the opposite of −5, which is 5. We could write that as – (–5)."

"How would I solve $-x = 5$?"

The response is usually silence.

I offer a hint: "I can write $-x$ as $-1x$. So now the equation looks like $-1x = 5$. What do I divide both sides by?"

I should hear "Divide both sides by –1." I tell them to do it. Most students will obtain $x = -5$.

Examples. I work with them on the first two.

1. $\frac{t}{3} = -11$. *Answer*: $t = -33$.

I may need to remind them that $\frac{t}{3} = \frac{1}{3}t$, and we multiply both sides by the reciprocal of $\frac{1}{3}$ which is 3.

2. $-6m + 2 = -7m$. *Answer*: $m + 2 = 0$, $m = -2$.

I advise adding $7m$ to each side to obtain a positive variable. This leaves 0 on the right-hand side. They then need to subtract 2 (i.e., add –2) from (to) each side.

3. $-7(r-1)=0$. *Answer*: *I advise that* $(r-1)=(r+(-1))$.

Distributing the –7 results in $-7r+7=0$; $-7r=-7$; $r=1$.

4. $7=\frac{n}{2}-1$. *Answer*: $\frac{n}{2}=8$; $n=16$.
5. Challenge problem: $x+36=1-4(x-5)$.

   Answer: $x+36=1-4x+20$; $5x=-36+21$; $5x=-15$, and $x=-3$.

Again, this involves distributing a negative number. A hint might be that the right-hand side can be written as $1+(-4)(x+(-5))$.

Literal Equations

"Suppose we had this problem to solve." $2x=10$.

"I know you know how to do this but before we do it, I'm going to write another on the board." I write: $ax=c$.

"I want someone to tell me what to do to solve the first one, step by step. First step is what?"

Someone will tell me to divide each side by 2. "Good. Now I'm going to do the equivalent on the second equation. What do I do to isolate x? I divide both sides by what?"

Students are hesitant but I will hear some say "Divide by a."

"OK. I get $x=\frac{c}{a}$. Remember we don't use the divide sign here; when dividing we use the fraction bar. This type of equation where we solve for another letter in the equation is called a 'literal equation.' The rules that apply to equations with numbers apply to literal equations"

Examples. I work through a number of these with the class.

1. Solve for x. $3x=a$. *Answer*: $x=\frac{c}{3}$ or $\frac{1}{3}a$.
2. Solve for x. $\frac{x}{2}=b$. *Answer*: $x=2b$.
3. Solve for t. $I=prt$ *Answer*: $t=\frac{I}{pr}$.

I give them a hint: "Think of pr as one number, and it is the coefficient of t. What do we do to isolate t?"

4. Solve for *b*: $p = a + b + c$. *Answer*: $b = p - a - c$.
5. Solve for *g*: $d = \frac{1}{2} gt^2$. *Answer*: $g = \frac{2d}{t^2}$.

Prompt: "First, what do we do to remove the denominator?"

Reponse: Multiply by 2.

"How do we now isolate g?"

Response: Divide by t^2.

Dividing by a variable squared may seem to students like something not allowed; I assure them it's legal. "It represents a number just like t represents a number, and as long as t isn't zero, we can divide both sides by it."

6. Solve for *r*. $c = 2\pi r$. *Answer*: $\frac{c}{2\pi}$.

Prompt: "What's the coefficient of *r*?"

HOMEWORK

Problems should be a mix of regular equations, which contain negatives and fractions like $\frac{2y}{3} + 7 = 5$, and literal equations. The latter should be kept fairly simple but have one or two challenging problems that can be worked through the next day, such as:

$V = \frac{1}{3} \pi r^2 h$; *solve for h.*

4. SOLVING ABSOLUTE VALUE INEQUALITIES

This lesson is preceded by working with inequalities and also solving absolute value equations. Students have learned how to solve inequalities in seventh grade. In a prior lesson (not included in the book) they learn to simplify inequalities such as $-2 < x + 4 \leq 5$. These inequalities are simplified by subtracting 4 from each side resulting in $-6 < x \leq 1$. Finally, students are to go the opposite direction and express $-6 < x \leq 1$ as two inequalities; i.e., $x > -6$ and $x \leq 1$. This amounts to writing one of the inequalities backwards which results in a reversal of the inequality sign. Knowing how to do this allows students to later write inequalities so that the variable is on the left-hand side—a convention that results in ease of interpretation.

Students have also learned previously the formal definition of absolute value. In seventh grade, they learned it as the distance from zero on the number line. The formal definition follows:

For any real number x, the ***absolute value*** *of x is denoted by |x| and is defined as:*

$$|x| = x \text{ if } x \geq 0 \text{ and } -x \text{ if } x < 0$$

Students get confused by this until they've had a bit more experience. I explain that when using a variable instead of a number, if x is negative (e.g., -6) then its absolute value is its opposite, or $-x$ *(e.g,* $-(-6)$*, or 6.*

Equations like $|x| = 6$, have two solutions. The first is obtained by removing the brackets and leaving the right-hand side positive. The second removes the brackets and makes the right-hand side negative. For $|x| = 6$, $x = 6$ and -6.

This lesson extends the concept of absolute value equations into inequalities. They learn to solve inequalities such as $|x + 2| < 5$.

WARM-UPS

1. Solve. $|x - 15| = 15$.

   Answer: $x - 15 = 15$;,$x = 30$; $x - 15 = -15$, $x = 0$.

2. Write an equation and solve. David is 17 pounds lighter than Paul. Their total weight is 259 pounds. Find David's weight.

   Answer: *Let x = Paul's weight,then x – 17 = David's weight.*

   $x + x - 17 = 259$; $2x = 276$; $x = 138$, $x - 17 = 121$.

   OR, Let x = David's weight; then x + 17 = Paul's weight.

   $x + x + 17 = 259$; $2x = 242$; $x = 121$ and $x + 17 = 138$.

3. Solve. $x - 2 < 2x - 3$. *Answer*: $x > 1$.

4. Write an equation and solve. Peter has \$180 and Ryan has \$150. How much money must Peter give Ryan so that they each will have an equal amount of money?

   Answer: *Let x = amount of money Peter pays and Ryan receives; then* $180 - x = 150 + x$; $2x = 30$; $x = \$15$.

5. Simplify. $-4 < x + 5 < 16$. *Answer*: $-9 < x < 11$.

Problem 3 is solved by adding 2 to each side and subtracting $2x$ from each side to obtain $-x < -1$. Dividing by -1 on each side results in $x > 1$. Alternatively, students may add 3 to each side and subtract x from each side resulting in $1 < x$. Reversing the inequality: $x > 1$.

GETTING THE WRONG ANSWER

I write on the board: $|x| < 5$.

"Go ahead and solve this. Let's see what you get." Students will solve it the same way as an absolute value equality, obtaining $x < 5$ and $x < -5$.

"Let's see if that's right," I say.

The first inequality works with a few test numbers less than 5 plugged into the left-hand side. For the second inequality, I ask for some numbers less than -5. Trying -6 results in $|-6|$ on the left-hand side, which equals 6. "Is 6 less than 5?" The answer is a resounding "no."

"In this case, we have to use the formal definition of absolute value which, if you recall, says that if $x < 0$, then $|x| = -x$. So, let's see what happens if we make the left-hand side negative." I write: $-x < 5$.

"What's next?" I'll ask and will hear that I should divide both sides by -1. The end result is $x > -5$, noting that division by a negative reverses the inequality sign.

"This makes more sense. So, we have $x < 5$ and $x > -5$. Let's graph it" (see Figure 4-1). (They know how to graph inequalities from a previous lesson.)

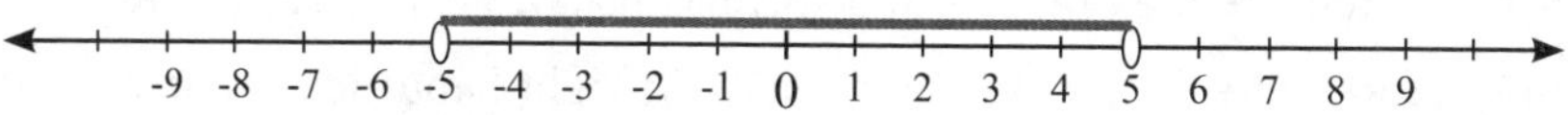

Figure 4-1

I'll then nudge the class to tell me how we can write this as one inequality statement rather than two, which happens to be: $-5 < x < 5$.

"So, this inequality represents all numbers that are between -5 and 5, but does not include either endpoint."

"How about $|n - 15| < 3$?"

They will get $n < 18$ for the first solution. For the second solution, they make the left-hand side negative: $-(n - 15) < 3$.

"What next? How did we solve the last problem?" They recall that both sides are divided by -1. $n - 15 > -3$, $n > 12$.

Examples

1. $|x| < 15$ *Answer*: $x < 15$ *and* $x > -15$.

2. $|x - 6| < 6$ *Answer*: $n < 12$ and $x > 0$.

3. $|y - 33| < -5$.

Some students will get this immediately—there is no solution because the absolute value of any number cannot be negative.

A Shortcut

"You are probably saying that there must be a quicker way to do these. And there is." There is a palpable sense of relief in the room as I introduce this quicker way based on a suggestion from a student in one of my algebra classes.

Looking at $|x - 6| < 12$, we see immediately that the first solution is $x < 18$. For the second solution, my former student said, "Since we're going to end up dividing by a negative, can't we skip the step of making the left-hand side negative, and just make the right-hand side negative and flip the inequality sign?" Summarizing the process in steps, I write:

Step 1: Remove the brackets and solve as usual. $x - 6 < 12$; $x < 18$.

Step 2: Remove the brackets and make the right-hand side negative, and reverse the inequality sign. $x - 6 > -12$.

Step 3: Solve the inequality. $x > -6$.

The solution is then $x < 18, x > -6$ or $-6 < x < 18$.

"Or" Situations

Now, I introduce problems for which the solutions are not common to both inequalities. These are associated with inequalities in the form $|x + c| > k$ in which c and k are constants. I write on the board: $|x - 3| \geq 5$ and have the students solve it.

The solutions are $x \geq 8$ and $x \leq -2$. We graph it and see that the solutions are not common to any particular interval (see Figure 4-2).

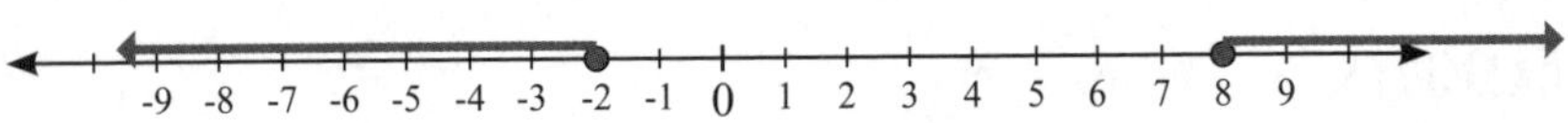

Figure 4-2

"In this situation we cannot write this as a single inequality because the numbers satisfying it are either in one interval or the other. This is an 'or' situation and we write it like this." $x \geq 8$ or $x \leq -2$.

To summarize this as a rule I say that when an inequality has a greater than sign, it's an "or" situation. When the inequality has a less than sign, it's an "and" situation.

I now announce a "challenge problem." I'll have them work this in their notebooks as I give hints and prompts loud enough for the whole class to hear—a technique that has benefits for all without embarrassment.

$$4|2x - 5| + 1 > 29$$

First prompt: "Is it an 'and' or an 'or' inequality?" I'm told it's "or."

Second prompt: "We have to isolate the absolute value term and its coefficient."

They now should have: $4|2x - 5| > 28$. Dividing by 4 yields: $|2x - 5| > 28$.

First solution: $2x - 5 > 7$, $x > 6$.

Second solution: $2x - 5 < -7$, $x < -1$.

I ask students to graph it, so they can clearly see why it is an "or" situation.

Homework

Other examples follow, some of which are from the homework. The homework consists of a combination of making single inequality statements from two inequalities and solving absolute value inequalities and equations.

5. UNIFORM MOTION PROBLEMS: OPPOSITE DIRECTIONS

We now progress into problems that involve uniform motion, which many students find difficult. Some of the difficulty comes from the wordiness of the problems—students need to identify what the problem is asking. My aim is to show the basic structures of these problem types and how they translate into equations. All uniform motion problems at this point in the course take the form of *distance* = *distance*. Later, when they have worked with algebraic fractions (rational expressions), such problems take the form of *time* = *time*.

This section addresses motion in opposite directions (i.e., two objects heading toward each other, or away from each other). The next section addresses objects moving in the same direction. A third category, round trips, is addressed later in the section on algebraic fractions, when such problems can be solved in the form *time* = *time*.

WARM-UPS

1. There are 900 students at Wayne High School. There are 20 more girls than boys. How many girls are there? *Answer* 460 *girls*: *Let* x *number of boys; then* $x + 20$ = *number of girls.* $x + (x + 20) = 900$; $2x = 880$; $x = 440$, $x + 20 = 460$.
2. A car traveled on a freeway for 2 hours at 80 mi/hr. It then traveled for another 3 hours at 70 mi/hr. How far did it travel in all? (*Hint*: Distance = rate × time.) *Answer*: $80 \times 2 + 70 \times 3 = 160 + 210 = 370$ *miles*.

3. A motorcyclist took 7 hours to travel between Town X and Town Y at an average speed of 35 mi/hr. A car took 5 hours for the same trip. Find the speed of the car. (*Hint*: Speed = Distance/time,)

   Answer: *Motorcycle travels* $7 \times 35 = 245$ *miles. Speed of car*: $\frac{245}{5} = 49$ *mi/hr.*

4. Solve the inequality; indicate whether it is "or" or "and." $|x - 45| > 5$.
   Answer: $x - 45 > 5$; $x > 50$, *OR* $x - 45 < -5$; $x < 40$.

5. Solve and write as one inequality rather than two. $|x + 3| < 4$.

   Answer: *First answer*: $x < 1$; *Second answer*: $x + 3 > -4$; $x > -7$; $-7 < x < 1$.

Students will likely need guidance for Problems 2 and 3. The distance = rate × time formula applies. Prompt for Problem 2: "How do you find its distance for the first part? How far does it travel in 1 hour going 80 mi/hr?" For Problem 3: Distance divided by total time to travel the distance = average speed. Prompt: "How far does the motorcycle travel in 7 hours?"

Distance = Rate × Time

Problems 2 and 3 of the Warm-Ups relied on the formula $D = rt$. I make a point of saying that 60 mph is a unit rate and means 60 miles per 1 hour.

"We will be working on uniform rate problems—that is, problems in which the objects are going at a steady speed. In all such problems, the solution relies on the formula 'distance = rate times time.'" I project a problem on the board:

Two cars head toward each other on the same road. They start at the same time. One starts from the north and travels south on this road at 70 mph. The other starts from the south and travels north at 80 mph. They meet at a certain point on that road. How far apart were they 1 hour before they meet on that road?

Some students get the answer right away, whereas others can't see how the problem is solved. I often hear, "There's not enough information," to which I reply, "Yes, there is."

I draw a vertical line and put a dot somewhere near the middle. "The dot is where the two cars meet. Can you tell me how far away from this dot the car going south is located one hour prior to getting there?"

Next, I prompt if I hear nothing: "How fast is the southbound car going?" They answer is 70 mph. "So, how far does the car travel in one hour?"

This generally results in a collective "OHHH."

One hour previous to the cars meeting, the southbound car is 70 mi north of the meeting point. The northbound car is 80 mi south of that point; 1 hour before they meet, they are 150 miles apart.

Some Problems

"Today we will be solving problems in which, like the problem I just gave you, there are two objects—cars, ships, bikes, people—traveling in opposite directions. They start at the same time, travel at a uniform speed, and are either heading towards each other, or away from each other."

Don rides at 15 mph, and Nancy rides at 12 mph. They go in opposite directions but start from the same point. How long until they are 135 mi apart?

"This is an opposite-direction problem. They are starting from the same place and heading away from each other."

I draw a picture to illustrate (see Figure 5-1):

Figure 5-1

"The total length of the line is 135 miles—that's how far apart they are after some amount of time. What are we trying to find?"

Students easily see we are trying to find the time it takes to be 135 miles apart.

"If I find the sum of the distances that Don and Nancy travel in that time, what is that total?"

If there is any hesitation, I tell them to look at the problem again. They see that the total distance is 135 miles.

I write: *Don's distance + Nancy's distance = Total distance (135 miles)*"

"For purposes of illustration and only for such purposes, let's use a guess and check approach."

I draw a chart on the board and fill it in with them. "Distance = rate × time. In one hour how far does Don go, riding at 15 mi/hr? And how far does Nancy go at 12 mi/hr?"

Time	**Don**	**Nancy**	**Total**	**= 135?**
1 hour	$15 \cdot 1 = 15$	$12 \cdot 1 = 12$	27	No
2 hours	30	24	54	No
3 hours	45	36	81	No

"In 1 hour, they go 15 and 12 miles, respectively, for a sum of 27 miles," I say. "Is 27 equal to 135? No, so 1 hour is not the answer to the problem. We do the same for 2 and 3."

"Now let's use algebra." I ask them to copy this chart:

	Rate × Time = Distance		
Don			
Nancy			

"Instead of plugging in numbers for the time they have travelled and trying them out to see if they add up to the right number, what do we use when we don't know a number in an equation?"

I usually hear "*x*" which is fine, although I like to use *t* just as a reminder that the variable represents time. "Let's fill out the chart. We know how fast Don and Nancy are going. But we don't know for how long. What do we put in the 'time' column?"

Response: *x*.

And how do I write 15 times *x*?" Continuing in this fashion the completed chart looks like this:

	Rate × Time = Distance		
Don	15	x	$15x$
Nancy	12	x	$12x$

"We now have $15x$ representing Don's distance after x hours and $12x$ is Nancy's.

Plugging in these terms into *Don's distance + Nancy's distance = 135*, we obtain the following equation:

$$15x + 12x = 135;\ 27x = 135;\ x = 5 \text{ hr}$$

Examples. I work with them, diminishing my support as they gain confidence.

1. Two hikers are 14 miles apart. The hikers start walking towards each other at 7 a.m.. One hiker is walking at 4 mph, and the other at 3 mph. How long will it take to meet?

"In this problem, they are walking toward each other. They start at the same time, and each walks at a constant rate of speed. Would you say that the time it takes for them to meet is the same for each person?"

They think about this for a moment and agree.

"So, what we have is this situation" (see Figure 5-2):_

Figure 5-2

"Each arrow represents a hiker. This is another 'distance = distance' problem." I write on the board: *First hiker's distance + Second hiker's distance = total distance (14 miles).*

"You can use a table like we did before. Let's see what you come up with."

Students work in their notebooks, and I go around offering help and guidance. The equation I want to see is: $3t + 4t = 14$. Solving it, $t = 2$ hours.

"Suppose I had asked you what time they will meet? How can we do that? Look at the problem." The hikers start at 7 a.m., students see that they meet at 9 a.m..

2. Two cars traveling in opposite directions meet on a highway. One averages 80 miles per hour and the other 70 miles per hour. In how many hours will they be 450 miles apart?

Prompt: "This is a 'traveling away from each other' problem." I refer them to the first problem we did that is similar. The equation for this example is $80t + 70t = 450$; $t = 3$ *hours.*

3. Two ships are sailing toward each other and are 120 nautical miles apart. If the rate of one ship is 4 knots greater than the rate of the other, and if they meet in 3 hours, find the rate of each ship.

In this problem, a table is useful, so I ask them to construct it. We are told they meet in 3 hours. "Do we know the rates of the ships?" No.

Some students look at this as if it's a betrayal. "Some problems ask for time, others for rate, and still others for distance. If one ship is going x knots, then how do we represent the speed of the other ship?"

Response: $x + 4$.

The equation is $3x + 3(x + 4) = 120$; $3x + 3x + 12 = 120$;

$6x = 108$; $x = 13$ knots, $x + 4 = 17$ knots.

Homework

I typically assign five or six problems, and I work on the first two with the class.

6. UNIFORM MOTION PROBLEMS: SAME DIRECTION

In this lesson, we look at problems with motion in same direction—for example, how long will it take a faster car to catch up with a slower one that had a head start of so many miles, or so many hours.

I use an approach in Foerster's (1990) *Algebra 1* textbook, which in my experience students find easier than other methods.

Today's Warm-Ups include a problem with motion in opposite directions, but also a problem in which there is motion in one direction at two different rates. The equation for this problem is similar to that used in the opposite direction problems, and it takes the form of Distance 1 + Distance 2 = Total distance. This problem is included to broaden the concept of "distance = distance" beyond that of opposite motion or today's "catch-up" type problems.

Homework problems and some of the examples in today's lesson include opposite motion as well as same direction so that problem solving ability is not limited to the latest thing students have learned.

As I indicated in the introduction to eighth-grade algebra, I am focusing on some of the key areas of algebra in this book, so there are topics that are omitted. With respect to word problems I have omitted discussion in this book of other problem types such as consecutive numbers and some types of mixture problems. That said, the Warm-Up problems may include problems that will have been addressed in an algebra course, even though omitted in this book.

WARM-UPS

1. Two planes start toward each other at the same times at airports that are 1,380 miles apart. One plane travels at 320 mph and the other at 600 mph. In how many hours will they meet? *Answer*: $320t + 600\ t = 1380$; $920t{=}1380$; $t = \frac{3}{2}$ *or* $1\frac{1}{2}$ *hr.*
2. In Problem 1, how far will the slower plane have traveled before meeting the other plane? *Answer*: $d = rt$; $d = 320 \cdot \left(\frac{3}{2}\right) = 480$ *miles.*
3. Two trains travel toward each other and are 388 miles apart. One train averages 47 mph. It takes 4 hours for the trains to meet. What is the speed of the other train? *Answer*: $47(4) + 4x = 388$; $188 + 4t = 388$; $4t = 200$; $t = 50$ *mph.*
4. Mr. Smith travels 665 miles, half of the time by car at 45 mph and half the time by train at 50 mph. How long did the trip take?

   Answer: *Let t = time for each half of the trip*; $45t + 50t = 665$;

   $95t = 665$; $t = 7$ *hrs*; entire trip takes $2t = 14$ *hr.*
5. Find four consecutive even numbers whose sum is 100.

   Answer: *Let* x *= first number. Then,the next three numbers are* $x + 2$, $x + 4$, *and* $x + 6$.

   $x + x +2 + x + 4 + x + 6 = 100$; $4x + 12 = 100$; $4x = 88$;

   $x = 22$, $x + 2 = 24$, $x + 4 = 26$, $x + 6 = 28$.

Problem 2 is an extension of Problem 1 and requires students to use the $d = rt$ formula. Problem 3 gives the time it takes for the trains to meet so that students are solving for speed. Possible prompt: "We have distance of one train + distance of the other train = 388. How do we represent the distance traveled by the train going 47 mph?" Problem 4 defines a trip that is split into two halves by time, so that t is the same for both halves. For Problem 5, as mentioned earlier, we have covered consecutive problems. Consecutive even numbers advance by 2 (e.g., 2, 4, 6).

RELATIVE SPEEDS

"Yesterday, we talked about motion in opposite directions. Today, we'll be talking about motion in the same direction."

I present a situation with which all students are familiar.

"You've been in a car that is passing another, and when you look at the car being passed, it appears that you're traveling at a slow speed relative to that car. So, if you pass the car going 65 mph and the car you're passing is going 60, what speed does it look like you're going?"

Response: 5 mph.

"We call this 'relative speed.' The speed of the faster car compared to the slower car is 5 mph. Now, suppose there is a car 10 miles ahead of you. You are going 70 mph, and the car ahead of you is going 60 mph. What is your speed relative to the car ahead of you?"

Response: 10 mph.

"Suppose you have telescopic vision and you watch the 10 mile space in front of you. It appears you are traveling the 10 miles at what speed?"

Response: 10 mph.

"What is our formula for distance that we've been using?"

Response: Distance = rate × time.

"How long it will take to go 10 miles at the speed of 10 mph—the speed it looks like we're going relative to the slower car. Tell me how you do it."

Generally the explanation comes back as "If you're going 10 mph, it takes 1 hour to go 10 miles."

"Correct," I say, "but I would like it stated as an equation. Distance between the cars is 10 mph. But instead of 70 for the rate, I'm going to use relative rate. Using $d = rt$, let's plug in the values that we know." I ask for someone to do this.

The equation is $10 = 10t$.

Solving Problems with Motion in Same Direction

"This principle is used to solve problems where we have motion in the same direction. Let's say we have a car that has gone 200 miles, and averages 60 miles per hour. How long would it take you to catch up to that car if you go after it at 100 miles per hour? Let's write out the steps."

I then write down the steps which they copy in their notebooks:

Step 1: Find the relative speeds.

What is the speed of your car at 100 mph relative to the 60 mph car? Answer: 100 – 60 = 40 mph.

Step 2: What is the gap in distance?

In this problem, the faster car must close the 200 mile distance between the two cars in order to catch up.

Step 3: Relative speed · time = gap in distance.

This is the form the equation will take. The faster car closes the gap of 200 miles at a relative speed of 40 mph: $40t = 200$.

Step 4: Solve the equation: $t = 5$ *hr*.

Examples. Problems are projected on the board. I ask students to work in their notebooks.

1. Mr. Smith starts driving from San Luis at 60 mph and has gone 120 miles when his wife starts after him at 80 mph. How long will it take Mrs. Smith to catch up to the Mr. Smith? *Answer*: *Relative speed of faster car compared to slower*:

   $80 - 60 = 20$ *mph*; *Gap in distance*: 120 *miles*; $20t = 120$; $t = 6$ *hr*

(I once had a student solve this the way I used to teach this type of problem. The distance that Mrs. Smith drives in catching up to Mr. Smith is equal to the 120 mile gap plus the distance Mr. Smith drives in the time t it takes for her to catch up: i.e., $60t + 120$. *Mrs. Smith's total distance is 80t.The equation is therefore 80t =* $60t + 120$, *which simplifies to the same equation used in the method above:* $20t = 120$. *This method was confusing to many students, so I used the relative speed method, which they generally find easier to understand.)*

2. Two airplanes start from the same airport at the same time and travel in opposite directions at 350 mph and 325 mph respectively. In how many hours will they be 2,025 miles apart? *Answer*: $350t + 325t = 2{,}025$; $675t = 2025$; $t = 3$ *hr*.

I ask the students, "Is this a catch-up problem or opposite direction problem? I tell them it is important to decide what type of problem it is before rushing in to solve.

3. A freight train left Beeville at 5 a.m. traveling 30 mph. At 7 a.m., an express train traveling 90 mph left the same station. When did the express overtake the freight? *Answer*: *Gap* = 30 *mph* · 2 *hrs* = 60 *mi. Relative speed* = 90 − 30 = 60; $60t = 60$; $t = 1$ *hr.*

Students say they don't know what the gap in distance is. I say, "Is there information in the problem that you can use to figure that out? How much time elapses between the time the freight train leaves and the express train starts out?"

Homework

Homework is a mix of same and opposite direction problems. In some problems, students need to find the gap distance; in others, students need to find the rate of one of the vehicles. In the time I leave time for students to start working on homework, I provide hints and guidance on the various problems.

7. POWER OF PRODUCTS AND OF POWERS

PRIOR CONTENT

After the unit on equations and introductory word problems, the next unit covers polynomials and operations. The unit includes adding/subtracting polynomials, power multiplication, multiplication of monomials and polynomials, and polynomial multiplication. In Part III, Chapter 2, the product of powers (e.g., $x^2 \cdot x^3$) was introduced. This was done to familiarize students with the procedure of multiplying powers to avoid overwhelming and confusing them with the full set of rules. The lesson previous to this one provides more complex multiplication of powers.

PRESENT LESSON

The previous lesson leads directly into the topic of taking the power of powers and the power of products. Specifically, students have learned that $(a^2)(a^2)(a^2)$ equals a^6 because the exponents are added. It then follows that for $(a^2)^3$ is the same as $a^{2 \cdot 3}$; that is, the exponents are multiplied. This is then extended to terms like $(a^2b^3)^2$ so that each of the exponents in the parentheses are multiplied by 2.

Students liken this to "distributing the exponent," and textbooks warn teachers to tell students not to characterize it in this manner. The reason is that by having students avoid using this description, it will help prevent them from assuming that they can apply the power to a power rule to binomials raised to a power; for example,

$$(a + b)^2 \neq (a^2 + b^2)$$

A friend who teaches high school math tells me she uses the word "apply" instead of distribute: "The exponent is *applied* to the numbers and variables inside the parentheses," and I agree that this is a good way to refer to it. I've found, however, that the distribution analogy for power of a product tends to help them see how the procedure works. That said, I use the distribution analogy with care; that is, I might say "When you apply the exponent to the numbers and variable, it's as if you are distributing it."

In my opinion and experience, telling students that the power of a product like $(a^2b^3)^2$ is not "distributing the 2 to the other exponents" is not likely to stop them from making the mistake with binomials no matter how many warnings are given.

Warm-Ups

1. $|2x + 4| \geq 5$. *Answer*: $2x + 4 \geq 5$; $2x \geq 1$; $x \geq \frac{1}{2}$;
   $2x + 4 \leq -5$; $2x \leq -9$; $x \leq -\frac{9}{2}$.
2. Two people start on bicycles for the same destination at the same time. The first person averages 15 mph and the second person 24 mph. In how many hours will they be 18 miles apart? *Answer*: *Let* t = *time for them to be* 18 *miles apart*;
   $24t - 15t = 18$; $9t = 18$; $t = 2$ *hr*.
3. $(x^3)(x^3)(x^3) = ?$ *Answer*: $x^{3+3+3} = x^9$.
4. $(2a^2)(2a^2) = ?$ *Answer*: $2 \cdot 2 \cdot a^2 \cdot a^2 = 4a^4$.
5. $h \cdot h^2 = ?$ *Answer*: h^3.

Problems 3, 4, and 5 will be used in the introduction to the day's lesson. Students will apply the rules for multiplying powers that they learned previously. For Problem 3, students apply the rule they learned in the previous lesson, summing the exponents. In Problem 4, students extend what they learned in the previous lesson, and multiply numbers together and then variables together. Problem 5 is included to remind students that a variable that does not have an exponent has an exponent of 1 that is not written. Therefore h can be expressed as h^1.

Power of a Power

I use the Warm-Ups to talk about the power of a power.

"Looking at Problem 3 of the Warm-Ups, we saw that we added the exponent 3 times." I write on the board: x^{3+3+3}.

"What if I had $m^2 \cdot m^2 \cdot m^2 \cdot m^2$? I can write this as $m^{2+2+2+2}$. Can I write this more compactly? Like, say, oh, I don't know, maybe, like a multiplication?"

I probably overdo the hint a bit, but they get where I'm going and tell me: $m^{2\cdot 4}$.

"And we see that m^2 is used as a factor four times, so we can also write it as $(m^2)^4$.

There is general agreement.

"What about $(x^3)^4$? How can I simplify this?"

Most see the pattern and tell me the exponents should be multiplied. With a few more examples to check for the basic understanding, I then write the rule for a power raised to a power:

For all positive integers m and n: $(b^m)^n = b^{mn}$.

I ask them to look at Problem 4 of the Warm-Ups. "You can see that we have the same thing going on here. The term $(2a^2)$ is used as a factor two times. So, we can write $(2a^2)\,(2a^2)$ in exponential form like this: $(2a^2)^2$. When you did Problem 4, you wrote it as $2 \cdot 2 \cdot a^2 \cdot a^2$. So I can write that as $2^2 \cdot (a^2)^2$."

I am writing this on the board, and they are writing this in their notebooks. This is a lot of information to take in, but we will be practicing this shortly on other problems.

"Each number and variable inside the parentheses is raised to the second power. Remember that 2 can be written as 2^1. We can then apply the rule for power of a power." I write on the board:

$$(2a^2)^2 = 2^{1\cdot 2}\, a^{2\cdot 2} = 2^2\, a^4$$

"We can write this as $4a^4$. Each number of variable inside the parentheses is raised to the power of 2. Which means that we multiply exponents as I've shown. Let's try one together:"

$$(3x)^4. \textit{ Answer: } 3^4\ x^4.$$

"We can simplify this as $81x^4$. What about this?"

$$(-3^x)^4. \textit{ Answer: } 81x^4.$$

"We are raising −3 to the fourth power. What if I had written it this way?"

$$-(3x)^4$$

I get a mix of answers, some $81x^4$ and others $-81x^4$. For those who get it right, I ask them to explain how they obtained their answer. I hear things like "The negative sign is outside the parentheses" or "The negative sign outside means the whole thing is negative."

More Examples.

1. $(ab)^2$. *Answer*: $a^2\ b^2$.
2. $(-2m^2)^3$. *Answer*: $-8m^6$.
3. $a^2 \cdot ab^3$. *Answer*: $a^3\ b^3$.

I indicate that this last example was not raising a power to a power, but the answer can be expressed in terms of a power to a power. I pose this as a challenge. *Answer*: $(ab)^3$.

4. $(-4w^7\ v^3)^3$. *Answer*: $-64w^{21}\ v^9$.
5. $-3x(4xy)^2$. *Answer*: $-3x(16x^2\ y^2) = -48x^3\ y^2$.

This one I work out with the students. "Remember what PEMDAS is? We treat exponents first, so we raise $4xy$ to the second power first and then multiply the result by $-3x$."

A common mistake is for students to multiply the –3 by 16 and also by x^2 and y^2 to obtain:

$-3 \cdot 16 \cdot -3 \cdot x^2 \cdot -3 \cdot y^2$.

This is a mistake, and I use the following example to show why:

"Suppose I had $2 \cdot (3 \cdot 4)$. Would I multiply the 2 by the 3 *and* the 4 like this: $2 \cdot 3 \cdot 2 \cdot 4$? No, you just multiply once. If I have $2(3x)$ what is the answer?" I should hear $6x$.

6. $b(-3b)^3$ *Answer*: $b(-27b^3) = -27b^4$.

Another common mistake: Students mistake the negative sign inside the parentheses as subtraction, and treat the problem as a distribution. Some mistakes include $b - 27b^3$ and others that make little sense. To prevent this I advise them to write things out as a string of multiplications: $(b) \cdot (-27) \cdot (b^3) = -27 \cdot b \cdot b^3$.

7. $(-r)(-6r)^3$ *Answer*: $-r(-216r^3) = 216r^4$.

For this, I remind them that $-r$ is the same as $-1r$, so we will have: $(-1) \cdot (-216) \cdot r \cdot r^3$. Since there are two negative numbers being multiplied, the answer will be positive.

Homework

I work with students on the homework problems, which are a mix of problems from yesterday's and today's lessons. Because this topic is confusing to students, I devote an extra day for additional explanation and practice as needed.

A subsequent lesson not included in this book covers quotients of powers and powers of quotients, such as $\left(\frac{a^2}{b^3}\right)^3 = \frac{a^6}{b^9}$.

8. MULTIPLYING TWO POLYNOMIALS

Students have learned how to use the product of powers to multiply monomials by polynomials in a previous lesson. This allows them to multiply expressions such as $2x(3x + 2x^2)$.

The multiplication of monomials and polynomials is an important skill in algebra and subsequent math courses. Multiplication of two polynomials is an extension of that skill. Students believe that it is going to be a difficult procedure. In fact, students find it quite easy and often anticipate the method before I finish my presentation of its genesis.

I do not present the shortcut for multiplying two binomials (referred to as FOIL) in this lesson. I wait until later. Because multiplication of polynomials other than binomials do not lend themselves easily to the FOIL technique, it is important that students learn the general method.

Students may still be making common mistakes, such as $x^2 + x^2 = x^4$ or $x^2 + x^2 = 2x^4$, as well as $x^2 + x^3 = x^5$. I therefore include problems in Warm-Ups to flush out such errors and disabuse students of these bad habits.

Warm-Ups

1. Simplify. $5x^2 + 3x^3 - 2x^2$. *Answer*: $3x^2 + 3x^3$.
2. Distribute. $(x + 2)a$. *Answer*: $ax + 2a$.
3. $x^2 + x^2 = ?$ *Answer* $2x^2$.
4. An airplane flew at x mph for 2 hours, then at $(x + 25)$ mph for another 4 hours. The entire distance was 1,150 miles. Find its rate (speed) during the 4-hour period.

   Answer: $2x + 4(x + 25) = 1150$; $2x + 4x + 100=1150$; $6x=1050$;

   $x = 175$ *mph. Four-hour period*: $(x + 25) = 200$ *mph.*

5. Solve and graph. $|3x - 6| \geq 18$.

   Answer: $3x - 6 \geq 18, x \geq 8$; $3x - 6 \leq -18, x \leq -4$.

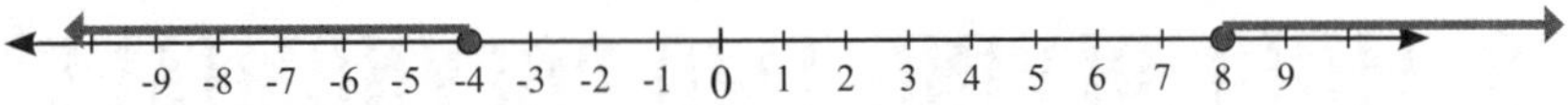

In Problems 1 and 3, some students may add exponents. Like terms may be combined by adding or subtracting, but exponents remain as is. For Problem 2, students may need to be told that a can be distributed just the same as if it were on the left-hand side of the parentheses. For Problem 4, possible prompts would be to write on the board: *Distance for 2 hours + distance for 4 hours = 1150*. "What's the formula for distance when we know rate and time?" "What is the time for the first leg of the trip? For the second? How do we represent speed for the first leg? For the second?" Problem 5 is an "or" situation; there is no number that satisfies both inequalities.

MULTIPLYING TWO BINOMIALS

By virtue of Problem 2 of the Warm-Ups (and some of the problems in their homework), students know that the number or variable to be distributed can be on either side of the parentheses. I use this in my explanation of the procedure for multiplying two binomials.

"In Problem 2 of the Warm-Ups, we had $(x + 2)a$. And you solved it just fine. Let's do it again, with a different binomial."

$$(2x + 3)z$$

"Now, distribute it." If all goes well, and it usually does, they will get this.

$$2xz + 3z$$

"We're going to learn how to multiply two polynomials, starting with two binomials. Let's start with $(2x + 3)(4x + 5)$."

I pause for dramatic effect, and they wait as if I'm a magician about to reveal an ancient secret.

"We distributed z across $(2x + 3)$, and the result was $2xz + 3z$. Now, let's let z equal $(4x + 5)$. I'm going to substitute $(4x + 5)$ for z in $(2xz + 3z)$. I'll do the first one. $2x\ (4x + 5)$. Now, you do $3z$."

Most students get $3(4x + 5)$. "Now, let's put it together."

$$2x(4x + 5) + 3(4x + 5)$$

"Now, distribute each one and combine like terms. Do it in your notebooks."

$$8x^2 + 10x + 12x + 15 = 8x^2 + 22x + 15$$

"What you've just done is distributed the binomial $(4x + 5)$ across $(2x + 3)$, just like you distributed the z across $(2x + 3)$. Remember, z is a variable; it represents a number. We can think of $(4x + 5)$ in the same way. It's a number, which we can distribute. Let's see this with another example."

$$(x + 7)(x + 2)$$

"If we think of $(x + 2)$ like z, we can distribute it to the x and the 7. Let's think of $(x + 2)$ as z first. Then, we'd get this."

$$(x + 7)z = xz + 7z$$

"Let's plug in $(x + 2)$ for z now."

$$x(x + 2) + 7(x + 2) = x^2 + 2x + 7x + 14 = x^2 + 9x +14$$

Treating an expression with two or more terms as a single entity is called "chunking," just as we think of an area code of a phone number is a single entity, rather than memorizing each digit of the area code as we would a new phone number. By now, most students are chunking the $(x + 2)$ binomial and seeing that it's treated as a single variable which is then distributed. I have them write the rule:

To multiply one polynomial by another, use the distributive property: multiply each term of one polynomial by each term of the other, and then add the products.

Examples

1. $(a + 8)(a - 7)$. *Answer*: $a^2 + a - 56$.

Sometimes someone will ask whether the $(a + 8)$ binomial can be "chunked" and distributed across the $(a - 7)$ binomial, and the answer is yes. It can be done either way.

2. $(y\ 9)(y + 5)$. *Answer*: $y^2 - 9y + 5y - 45 = y^2 - 4y - 45$.
3. $(n - 9)(n + 4)$. *Answer*: $n^2 - 9n + 4n - 36 = n^2 - 5n - 36$.

Multiplying a Binomial by a Trinomial

The same method applies when multiplying a binomial and a trinomial. I write on the board: $(x + 5)(x^2 - 3x - 2)$.

"In this case, what do we 'chunk' and how do we distribute it?"

Some students will say to distribute the trinomial across $(x + 5)$, whereas most will say to distribute the $(x + 5)$ across the trinomial. I'll have each student write it out at the board to see that the answers are the same. I check carefully as they do it to ensure there are no mistakes along the way so the result is in fact the same.

$$(x + 5)(x^2 - 3x - 2) = x(x^2 - 3x - 2) + 5(x^2 - 3x - 2) =$$
$$x^3 - 3x^2 - 2x + 5x^2 - 15x - 10 = x^3 + 2x^2 - 17x - 10$$

Examples

1. $(x + 1)(x^2 + 5x + 8)$. *Answer*: $x^3 + 6x^2 + 13x + 8$.

I offer the following suggestion. When combining terms, it helps to write the terms in two rows, with like terms underneath each other to make adding them easier:

$$\begin{array}{r} x^3 + 5x^2 + 8x \phantom{{}+8} \\ +\quad x^2 + 5x + 8 \\ \hline x^3 + 6x^2 + 13x + 8 \end{array}$$

2. $(n - 3)(n^2 - 2n + 4)$. *Answer*: $n^3 - 5n^2 + 10n - 12$.

I will show maybe two more examples and then start students on the homework.

Homework

Problems will consist mostly of multiplying two binomials but will also include binomial times a trinomial and a challenge problem of two trinomials. Also included are problems where a monomial is multiplied by a binomial which they learned in the previous lesson.

9. ZERO POWER AND NEGATIVE EXPONENTS

PRIOR CONTENT

Students have learned to work with quotient of powers:

For positive integers m and n, and $b \neq 0$, if $m > n$ then $\frac{b^m}{b^n} = b^{m-n}$.

For positive integers m and n, and $b \neq 0$, if $m < n$ then $\frac{b^m}{b^n} = \frac{1}{b^{n-m}}$.

They have also learned the rule for powers of quotients: $\left(\frac{a^x}{b^y}\right)^n = \left(\frac{a^{xn}}{b^{yn}}\right)$.

PRESENT LESSON

For quotients such as $\frac{b^2}{b^9}$, students have learned to "subtract up" so that the expression is $\frac{1}{b^{9-2}}$ or $\frac{1}{b^7}$.

When teaching students about this, I inevitably get asked, "Isn't it also "b^{-7}?" I answer that it is, but for now we'll do it this way, and later we will learn negative exponents. This is that later lesson.

I use the rules of exponents to show what negative exponents represent, as well as what the zero power represents. I have seen approaches in which both the zero power and negative exponents are explained in terms of powers of 10. Specifically, 10^3 is $\frac{1}{10}$ of 10^4, 10^2 is $\frac{1}{10}$ of 10^3 and so on. When we get to 10^1 (i.e., 10), then the pattern tells us that 10^0 must be $\frac{1}{10}$ of 10 or 1. When we go further to the right on this power of 10 number line, then 10^{-1} is $\frac{1}{10}$ of 1, and so on. Having used both methods, I've found that using the exponent rules to explain zero and negative exponents is easier for students to understand.

Another approach I've seen for the zero power is showing that x^3, say, is a number in which x is a factor 3 times—but multiplied by 1: that is, 1·3·3·3. Then, 3^1 is 1·3, and $3^0 = 1$, since 3 is used as a factor 0 times. I will sometimes refer to this explanation when students try to articulate what a zero power represents.

WARM-UPS

1. Divide. $\frac{x^5}{x^8c^4}$. *Answer*: $\frac{1}{x^3c^4}$.
2. Simplify. $\left(\frac{ab^2}{3}\right)^2$. *Answer*: $\frac{a^2b^4}{9}$.
3. Solve. $\frac{2}{3}x = 6$. *Answer*: $x = 6 \cdot \frac{3}{2} = 9$.
4. $\frac{1}{\frac{a}{b}} = ?$ *Answer*: $\frac{b}{a}$.
5. Simplify. (Divide first.) $\frac{3x^4}{x^3} + \frac{4x^2}{x}$. *Answer*: $3x + 4x = 7x$.

These problems go over the powers of quotients rule. Problem 4 reinforces the idea that 1 divided by any number is the reciprocal of that number. Thus, the reciprocal of $\frac{a}{b}$ is $\frac{1}{\frac{a}{b}} = 1 \div \frac{a}{b} = 1\times \frac{b}{a} = \frac{b}{a}$.

ZERO EXPONENT

"What do you think 5^0 is?" I ask my students, who almost always respond "Zero!" and to which I reply, "Let's look at this more closely."

I write on the board: $\frac{5^4}{5^4}$

After getting agreement that any number divided by itself is 1, I then ask how we would use the rule of exponents for quotients, reminding them that the rule calls for subtracting exponents. They quickly see that it is equal to 5^{4-4} or 5^0.

"We now have the following: $5^0 = \frac{5^4}{5^4}$, and $\frac{5^4}{5^4} = 1$. If I apply the rule of transitivity, what can we say about 5^0?"

The light goes on at this point, although I've had some students ask what the rule of transitivity is. I have to remind them. (Forgetting is a normal part of math classes, which is why we have to keep repeating things they've learned.)

I write the following: $b^0 = 1$ *for every* $b \neq 0$.

"You may be wondering why I say that b cannot equal 0. Let's take a look. If I have 0 to any power, the answer is 0. So, let's see what happens if I have this:

$$\frac{0^5}{0^5}$$

"We all agree that this is $\frac{0}{0}$, but if you recall from last year, what did we say about dividing 0 by 0?"

Someone usually will say, "It equals 1." And I will respond, "Can it equal 2? How about 5? What is 0 times any number?" I then remind them that 0 divided by 0 is "indeterminate" since there is no single number that satisfies it.

Examples. These serve to expand how the zero power is used in expressions.

1. 1000^0. *Answer*: 1.
2. $(-3)^0$. *Answer*: 1.
3. $\frac{x^6}{x^0}$. *Answer*: $\frac{x^6}{1} = x^6$ or $\frac{x^6}{x^0} = x^{6-0} = x^6$.
4. $(7b)^0$. *Answer*: 1.
5. $7x^0$. *Answer*: $7 \cdot 1 = 7$.

Students see that it is only x being raised to the zero power; the 7 is multiplied by x^0, whereas in Problem 4, the whole expression of $7b$ is raised to the zero power.

6. $-8f^0$. *Answer*: -8.

Negative Exponents

"Somebody simplify these for me; write it on your mini-whiteboards."

$$\frac{a^7}{a^3}, \frac{a^3}{a^7}$$

They've been doing problems like this so they come up with the answers fairly quickly: a^4 and $\frac{1}{a^4}$.

"Some of you asked if we could subtract exponents and express the answer as a^{-4}. And as a matter of fact, you can. Before I get further into that, bear with me. Problem 4 of the Warm-Ups had this complex fraction: $\frac{1}{\frac{a}{b}}$. When we divide 1 by any number, the result is called what?" We have gone over this in the Warm-Ups, so they haven't yet forgotten that it's the reciprocal.

"What's the reciprocal of $\frac{3}{2}$?" Hearing $\frac{2}{3}$, I then ask what is the reciprocal of $\frac{a^3}{a^7}$.

Some will respond a^4, and others will respond $\frac{1}{\frac{a^3}{a^7}}$ and still others $\frac{a^7}{a^3}$. "Good; all are correct, so we'll work with all of them." I write these on the board: $\frac{a^7}{a^3}$, $\frac{1}{\frac{a^3}{a^7}}$, $\frac{1}{\frac{a^7}{a^3}}$ and a^4. I then show how these are all related, walking students through the steps as I ask questions:

$$\frac{a^7}{a^3} = a^4 \textit{ and } \frac{a^7}{a^3} = \frac{1}{\frac{a^3}{a^7}}.$$

$$\frac{a^3}{a^7} = a^{-4} \text{ and } \frac{1}{\frac{a^7}{a^3}} = \frac{1}{a^4}$$

$$\textit{Therefore: } a^{-4} = \frac{a^3}{a^7} = \frac{1}{\frac{a^7}{a^3}} = \frac{1}{a^4}.$$

Although this makes logical and perfect sense to me, I realize that some students will get lost along the way. To keep students from thinking this is more complicated than it really is, I say, "The good news is that we don't have to go through all this when working with negative exponents. We just have to remember a simple rule." I then write the rule:

$$b^n = \frac{1}{b^{-n}} \text{ and } b^{-n} = \frac{1}{b^n}$$

"If I said I want to write $\frac{1}{a^{-4}}$ so that I have no negative exponents, I would write a^4. They are the same thing. How would I write $\frac{1}{x^{-3}}$ with a positive exponent?"

The answer of course is x^3.

Examples

1. Write with a negative exponent. $\frac{1}{n^2}$. *Answer*: n^{-2}.
2. Write with a positive exponent. x^{-1}. *Answer*: $\frac{1}{x}$.
3. Write with a positive exponent. $3x^{-1}$. *Answer*: $\frac{3}{x}$.

Again, this is a multiplication where only x is raised to a negative power. The 3 is not raised to a power; it is $3 \cdot x^{-1}$. The 3, therefore, stays as 3, and x gets "moved downstairs."

4. Write with positive exponents. $3x^{-2}y$. *Answer*: $\frac{3y}{x^2}$.

This can be written as $3 \cdot y \cdot x^{-2}$ to make it clearer.

5. Evaluate. 10^{-3}. *Answer*: $\frac{1}{10^3} = \frac{1}{1000}$ *or* 0.001.
6. Evaluate. $\left(\frac{2}{3}\right)^{-2}$.Answer: $\frac{1}{\left(\frac{2}{3}\right)^2} = \frac{1}{\frac{4}{9}} = \frac{9}{4}$.

This one I walk through with the students. It lends itself to a shortcut: a fraction raised to a negative power is the reciprocal of that fraction raised to a positive power. But I do not introduce that yet, since they are processing a lot of information now. In a later lesson, I will present it.

7. Write with positive exponents. $\frac{x^2}{z^{-7}}$ *Answer*: x^2z^4

Homework

The homework is a mix of zero and negative exponents, keeping them fairly straightforward as in the examples. More complex problems are introduced at the end as challenge problems, such as: Express with positive exponents:

$\frac{p^{-2}}{q^{-1}}$. *Answer*: $\frac{q}{p^2}$.

10. IDENTIFYING COMMON FACTORS

After lessons on polynomials (which includes multiplying and dividing powers), students then learn about factoring. Students learn to factor expression such as $2x^2 + 4x$, for which the greatest common factor is $2x$ and the factored expression is $2x(x + 2)$. Also covered are:

1. The product of the sum and difference of two variables: $(x - y)(x + y)$ which equals $x^2 - y^2$.
2. The squaring of binomials using $(a + b)^2 = a^2 + 2ab + b^2$, and the factoring of perfect trinomial squares such as $x^2 + 4x + 4$.
3. Factoring trinomials in the form $ax^2 + bx + c$, for $a = 1$, and the more general case when a is greater than 1.
4. Complete factorization in which an expression such as $2x^2 + 10x + 12$ is factored as $2(x + 3)(x + 2)$.
5. Solving quadratic equations such as $x^2 + 5x + 6 = 0$ by factoring.

Because of the amount of information presented, students can become confused over how various expressions are factored. What was once straightforward such as factoring $2x^2 + 4x$ gets left unfactored because students will have become used to factoring trinomials. Conversely, an equation like $x^2 + 5x + 6 = 0$ is factored as $x(x + 5) = -6$, and is left unsolved. Interleaving of the various types of factoring is therefore essential in the unit on factoring so that students are not left with mastery of the latest technique learned, with the others forgotten.

The lesson prior to this particular one addresses what factoring is, and identification of greatest common factors (GCFs). Students have learned about GCFs sometimes as early as fifth grade. The prior lesson expands this so students must find the GCF of two terms such as $5b$, and $40bc$. (Answer: $5b$.)

Most problems with factoring involve an intuitive approach to finding the GCF between two numbers such as 8 and 12 (for which it is 4). Students also know when the greatest common factor is the smaller of two numbers such as 4 and 12, and when the second number is a multiple of the first, such as 6 and 18.

WARM-UPS

1. $-24a^2b^3 = -8ab^2 \cdot$ __. (Fill in the blank.) *Answer*: $3b$.
2. Find the GCF between $60m^3n$ and $48m^2n$.

   Answer: $12m^2n$.
3. Write the expression with a negative exponent. $\frac{a}{b^2}$. *Answer*: ab^{-2}.
4. Multiply. $-5x^2\,(x^3y^2)^2$. *Answer*: $-5x^2(x^6y^4) = -5x^8y^4$.
5. Distribute. $x(a + b)$. *Answer*: $ax + bx$.

Problems 1 and 2 have been covered in the lesson previous to this one. Problems 3 and 4 are a review of working with exponential expressions. Problem 5 serves as a lead into today's lesson.

COMMON MONOMIAL FACTORS

Problem 5 of the Warm-Ups presents a distribution problem.

"In Problem 5, we had a distribution for which the answer is $ax + bx$. Today we'll learn to factor expressions like this, which involves working backward. That is, given $ax + bx$, we sometimes need to work backward to find out what it looked like before it was distributed. We already know the answer: $x(a + b)$. In this case, x is the common monomial factor of $ax + bx$."

"For $2x + 10$ this is a final distribution product. What was the number that was outside the parentheses before distribution?"

Students usually tell me the answer quickly: 2.

"Correct. How would I write what it looked like pre-distribution?" I have them write on their mini-whiteboards: $2(x + 5)$.

"What you're really doing is the opposite of what you do in distribution. To distribute the 2, you multiply it by the variables and numbers inside the parentheses. When we factor it, what are we doing?"

Response: Dividing.

"Yes. And by dividing 2x and 10 by the GCF of the binomial, we are undoing the distribution"

I ask them to write this definition in their notebooks:

The greatest common monomial factor of a polynomial is the common monomial factor having the greatest numerical coefficient and the greatest degree of the variable(s).

I ask students to paste into their notebooks the steps for factoring a polynomial:

$$6a^2b - 15ab^2$$

Step 1: Determine by inspection the greatest common factor. *3 is common to* 6 *and* 15, *and ab is common to* a^2b *and* ab^2; 3*ab is the GCF* since a and b are variables of the greatest degree (i.e., with the highest exponents) that are common to the variables in the expression.

Step 2: Write the greatest common factor outside a set of parentheses. $3ab(\ )$.

Step 3: Fill in the quantity within the parentheses by dividing each term of the polynomial by the greatest common factor. $3ab(2a - 5b)$.

Examples. These are taken from the first homework problems assigned and worked together as a class so they are getting a jump on the homework.

1. What is the GCF of $4a^2$ and $12a$? *Answer*: $4a$.

I ask why we can't factor out a^2. The technical answer is that a is the variable of the greatest degree by which both terms can be divided. The more typical answer is that a is the greatest common factor between a and a^2.

2. *Factor.* $5x^2 - 3x$. *Answer*: $x(5x - 3)$.
3. Factor. $6z^4 + 36z^3 + 60z^2$. *Answer*: $6z^2(z^2 + 6z + 10)$.
4. Factor. $x^2 - xy$. *Answer*: $x(x - y)$.
5. Factor. $14m^2 + 42m$. *Answer*: $14m(m + 3)$.
6. Factor. $10x^2y + 15xy - 5y$. *Answer*: $5y(2x^2 + 3x - 1)$.

7. Factor. $mn^2t - m^2nt^2$. *Answer*: $mnt(n - m)$.
8. Factor. $-21v^3w^2 + 14v^2w^5$. *Answer*: $7v^2w^2 (-3v + 2w^3)$.

HOMEWORK

The homework problems include binomials and trinomials all with common factors. Problems that require factoring by grouping are discussed in a subsequent lesson. That lesson is not included in this book; a summary of factoring by grouping follows below.

FACTORING BY GROUPING

Factoring by grouping involves chunking binomials. To factor $ax + by + ay + bx$ we first group it into $(ax + bx) + (ay + by)$. In this form, it can then be factored as follows::

$$(a + b)x + (a + b)y$$

The binomial $(a + b)$ is a common factor of the above expression. Students learn to think of a binomial like (a + b) as a single variable like T that can be factored out; for example, $xT + yT = T(x + y)$. In the case of the above expression, factoring out the binomial $(a + b)$ results in: $(a + b)(x + y)$.

11. SQUARING A BINOMIAL

PRIOR CONTENT

Prior to this lesson, students have learned a shortcut for multiplying the sum and difference of two quantities. Specifically, $(x + y)(x - y) = x^2 - y^2$. When multiplying it out using the method, they have learned that when multiplying polynomials they obtain $x^2 - xy + xy - y^2$.

The middle terms drop out since they are opposites. Working in reverse, given the difference of two squares, it can be factored into two binomials that are the sum and difference of the respective square roots of the two terms of the binomial. Students learn to identify the form of the difference of two squares.

Shortcut for Multiplying Two Binomials. After learning to multiply two polynomials students have learned the FOIL method of multiplying two binomials. There has been some criticism of the FOIL method as obscuring the conceptual underpinning of what is happening when multiplying binomials. I provide a derivation of the shortcut. The first method they learned, as discussed in Part III, Chapter 8, gives way to the shortcut. I admonish my students that this initial method is not to be forgotten because the shortcut is hard to apply for multiplying polynomials other than binomials; for example, $(a + b + c)(d + e + f)$.

I show the derivation of using the multiplication of $(a + b)(c + d)$, using the method they have been using:

$$(a + b)(c + d) = a(c + d) + b(c + d) = ac + ad + bc + bd$$

Each of the terms in the product can be related the position of each of the terms in the two binomials as follows:

ac = the product of the first terms in each of the two binomials: a and c.

ad = the product of the two outside terms of each of the two binomials: a and d.

bc = the product of the two inner terms of each of the two binomials: b and c.

bd = the product of the last terms of each of the two binomials: b and d.

This gives way to the mnemonic FOIL, which stands for "first, outer, inner, and last" as described above and provides an easier way to do the multiplication. I have found that students find it helpful when learning the factoring of trinomials, often remarking that it is a "reverse FOIL."

Current Lesson

Squaring a binomial provides yet another shortcut, and knowing FOIL helps us quickly get to it.

I remind students that it is to their benefit to know the squares of 11–20 and to be able to recognize numbers that are perfect squares. I post these on the board every day, but soon stop doing that. They will find problems on Warm-Ups are far easier if they have these squares committed to memory, since I don't allow them to use calculators. The chart I display is shown below:

11	12	13	14	15	16	17	18	19	20
121	144	169	196	225	256	289	324	361	400

Warm-Ups

1. Write in factored form. $289x^2 - 256y^2$. *Answer* $(17x - 16y)(17x + 16y)$.
2. Find the product. $\left(2y + \frac{1}{4}\right)\left(2y - \frac{1}{4}\right)$. *Answer* $4y^2 - \frac{1}{16}$.
3. Factor. $4x^3 - 16x^2$. *Answer*: $4x^2(x - 4)$.
4. Two cars leave town at the same time and go in the same direction. One car averages 60 mph and the other car averages 72 mph. How long will it take the cars to be 84 miles apart?

   Answer: *Let t =time for the cars to be* 84 *miles apart.*

   $72t - 60t = 84$; $12t = 84$; $t = 7$ *hours.*

5. Multiply. $(x + 2)(x + 2)$. *Answer*: $x^2 + 4x + 4$.

Problem 1 illustrates the advantage of knowing the squares from 11 to 20. Students may rely on the chart I put on the board. Problem 3 is not a difference of two squares, although some will treat it as such. It is important that students learn to identify what type of factoring will apply. Problem 4 is a distance = distance problem. In this case, the distance the faster car travels in t hours eventually will be 84 miles greater than the distance the slower car travels in the same amount of time. A diagram is helpful to students for such problems. Problem 5 provides a segue to the day's lesson. In going over the answer, I point out that the problem may also be written as $(x + 2)^2$.

Preliminaries

I write Problem 5 of the Warm-Ups with the answer on the board:

$$(x + 2)^2 = x^2 + 4x + 4$$

"Now, I want you to find $(x + 3)^2$." I may have to remind them that it is the same thing as $(x + 3)(x + 3)$.

They do so, and I have them square $(x + 4)$ and $(x + 5)$. I then have them look at the board which has the answers displayed.

$$(x + 2)^2 = x^2 + 4x + 4$$
$$(x + 3)^2 = x^2 + 6x + 9$$
$$(x + 4)^2 = x^2 + 8x + 16$$
$$(x + 5)^2 = x^2 + 10x + 25$$

"What patterns do you see?" I ask.

All notice right away that the last number is the square of the second term of the binomial. Others see that the coefficient of the middle term of the trinomial is two times the second term.

I then write the following problems on the board, have them do it:

$$(x - 2)^2 = x^2 - 4x + 4$$
$$(x - 3)^3 = x^2 - 6x + 9$$

"What is the pattern you see now?"

Generally, they see right away that the middle term is negative, and the last term remains positive.

"Does this suggest a rule that we can use for squaring a binomial?" Students may wrestle with this but ultimately come up with something that sounds close to "The first term is squared, and the second term is double the second term; the last term is the second term squared." This is right as far as the examples given, but is not the general rule; nevertheless, it gives me something to work with, and I ask them to do one last one: $(2x - 5)^2$. They come up with $4x^2 - 20x + 25$.

"This one follows the same pattern as the ones we just did. So, let me state the rule formally. It's a rule that is used to square binomials and can save time."

Rule for Squaring Binomials

I write the rule on the board using the last example: $(2x - 5)^2$.

Step 1: Square the first term in the binomial. $(2x)^2 = 4x^2$.

Step 2: Double the product of the two terms of the binomial.

The two terms are $2x$ and -5. The product is doubled: $2(2x)(-5) = -20x$.

Step 3: Square the second term in the binomial. $(-5)^2 = 25$.

Step 4: Combine the terms. $4x^2 - 20x + 25$.

Students may get confused when the second term is negative. I remind them that the binomial $2x - 5$ is the same thing as $2x + (-5)$.

Some students ask if they can use the FOIL method for squaring a binomial. I answer that they may use whatever method they feel comfortable with. I point out that the rule saves some time. It is also of value as a way to identify perfect square trinomials, and factor them. That topic is covered in a future lesson but is not included in this book.

Examples

1. $(m + h)^2$. *Answer*: $m^2 + 2mh + h^2$.
2. $(3a - 2)^2$. *Answer*: $9a^2 + 2(3a)(-2) + (-2)^2 = 9a^2 - 12a + 4$.

3. $(x + 9y)^2$. *Answer*: $x^2 + 18xy + 81y^2$.
4. $(ab - 2)^2$. *Answer*: $(ab)^2 - 4ab + 4$ or $a^2b^2 - 4ab + 4$.
5. $(4 + x^2)^2$. *Answer*: $16 + 8x^2 + x^4$.

Homework

The homework should primarily be problems that require squaring binomials some of which include terms that are powers such as in Problem 5 above. The assignment should also include other types of multiplication and factoring problems that have been covered in the past few lessons to ensure students are familiar with all these types.

12. GENERAL METHOD OF FACTORING TRINOMIALS

Previous to this lesson, students have learned how to recognize and factor perfect square trinomials such as $x^2 - 6x + 9$. They have also learned to factor non-perfect square trinomials in the form $ax^2 + bx + c$, where $a = 1$. The various cases to consider when factoring are:

1. All signs positive. Example: $x^2 + 5x + 6$; factored form: $(x + 3)(x + 2)$.
2. Middle term negative, last term positive. Example: $x^2 - 5x + 6$; factored form: $(x - 3)(x - 2)$.
3. Last term negative. Examples: a) $x^2 - x - 6$; factored form: $(x - 3)(x + 2)$. b) $x^2 + x - 6$; factored form: $(x + 3)(x - 2)$.

Students find the FOIL mechanism helpful in factoring trinomials, and think of it as a "reverse FOIL." For example, in factoring $x^2 + x - 12$, students reason that since the last term is negative, it must be the product of a negative and positive term. This leads to writing a partial solution: $(x + \_\_)(x - \_\_)$. The factors must sum to 1. Thus, the factors of 6 and –2 and 12 and –1 are eliminated, leaving –3 and 4 and –4 and 3. Of these, 4 and –3 sums to 1. The factored form is therefore $(x + 3)(x - 4)$.

Today's lesson covers the general form of factoring $\mathrm{A}x^2 + \mathrm{B}x + \mathrm{C}$ in which $\mathrm{A} > 1$, and A is an integer. These are more difficult for students since there are more possibilities to consider.

Warm-Ups

1. $(2x^3 - 4y)^2 = ?$ *Answer*: $(2x^3)^2 + (2)(2x^3)(-4y) + (-4y)^2 =$ $4x^6 - 16x^3y + 16y^2$.

2. Multiply. $2x^2(-4x^3y)^3$. *Answer*: $2x^2(-64x^9y^3) = -128x^{11}y^3$.
3. Write in factored form. $324x^6 - 196y^4$.
   Answer: $(18x^3 - 14y^2)(18x^3 + 14y^2)$.
4. Factor. $x^2 - 9x - 10$. *Answer* $(x - 10)(x + 1)$.
5. Solve. $\frac{1}{4}x - \frac{1}{5}x = 3$. *Answer*: *LCD* = 20. *Multiply each term by* 20:
   $20\left(\frac{1}{4}x\right) - 20\left(\frac{1}{5}x\right) = 20 \cdot 3$; $5x - 4x = 60$; $x = 60$.

Problem 1 requires the squaring of a power; also, the middle term is found by multiplying $2(2x^3)(-4y)$, which is $-16x^3y$. Problem 2 requires raising the term in parentheses to the third power first, and then multiplying by $2x^2$. Some students may view the negative sign in the parentheses as a subtraction sign and try to distribute; this problem should therefore be used as an example that it requires multiplication, not distribution. Problem 3 uses squares of 18 and 14. There should be no chart of squares on the board, so that students must rely on memorization. Problem 4 eludes some students because of the obvious factors of –10 being –10 and 1. Problem 5 revisits equations with fractions eliminating denominators by multiplying all terms on both sides by a common denominator; in this case, 20.

Factoring Trinomials When A and/or C are Prime Numbers

"So far we've been factoring trinomials where the first term has a coefficient of 1. Can you guess what we're going to do today?"

Most students figure out immediately where I'm going with this. "Let's look at the easy ones first. You've no doubt discovered that prime numbers like 3, 5, 11 and so forth only have one set of factors."

Saying it this way opens me up to someone pointing out that if the number is negative there are two sets of factors; for example, for –5 it is 5 and –1, and 1 and –5. I take note of such students as candidates for more challenging problems.

"Let's say we have these two problems to solve."

$$3x^2 + 2x - 1 \text{ and } 2x^2 - 3x - 9$$

"In the first problem, the leading coefficient is 3. There are only two factors of 3. What are they?" As the class shouts out 3 and 1, I write:

$$(3x\quad)(x\quad)$$

"You'll notice I didn't put in a plus or minus sign in the parentheses because that's my next step. What are the factors of –1 ?"

Response: –1 and 1.

"So, where do I put the –1 and where does the 1 go?" This is where "reverse FOIL" comes in handy, since they are finding the middle value by "outer product" + "inner product."

I have them work it out in their notebooks and they get the answer fairly quickly: $(3x - 1)(x + 1)$,

Students can see they need the outer product of $3x$ to be positive in order to obtain $2x$, which is the middle value of the trinomial.

"Let's try the second one: $2x^2 - 3x - 9$. In this case, the last number is *not* prime. Oh no! What're you going to do? Well, we know that 2 is prime so let's do the first step."

$$(2x\ )(x\ )$$

"Now, what are the possibilities for the factors of 9, given that the middle term must be negative?"

Response: 9 and −1, −9 and 1, 3 and –3.

"Let's try some combinations. Do either 9 and −1 or −9 and 1 seem like possibilities?"

They quickly see that the middle value will be too high or too low no matter where they put the 9 or −9. That leaves 3 and −3, which I have them work on by themselves. They quickly come up with the answer:

$$(2x + 3)(x - 3)$$

"Now, on a test or quiz if you ask me, 'Is this right?' I won't answer you, but how would you find out if you're right?"

Response: Multiply the binomials.

I give more examples, with increasing complexity. We do the first together and the rest they do independently.

EXAMPLES

1. $2x^2 + 5x + 2$. *Answer*: $(2x + 1)(x + 2)$.
2. $5a^2 + 4a - 1$. *Answer*: $(5a - 1)(a + 1)$.
3. $3x^2 - 10x + 8$. *Answer*: $(3x + 2)(x - 4)$.
4. $2y^2 + 7y + 3$. *Answer*: $(2y + 1)(y + 3)$.
5. $3x^2 + 20x - 7$. *Answer* $(3x - 1)(x + 7)$.

I throw in a non-factorable problem as well.

6. $3x^2 - 17x - 20$. *Answer*: *Not factorable.*

MORE DIFFICULT FACTORING

Trinomials for which A and/or C are *not* prime numbers are generally more difficult for students because of many possible combinations of factors. I generally devote a separate lesson for this but include it here.

I start by writing a problem on the board:

$$8x^2 - 14x + 3$$

"This problem is different from the ones we've been doing. The last number is a prime, but the leading coefficient is not. But before we get into that, notice that the last number is positive but the middle number is negative. What does that tell us?"

Response: Both numbers in the binomials will be negative.

"We will have something like this," I say: $(? - 3)(? - 1)$.

"Now, all we have to do is find the factors that go in the blanks. So, what are the factors of 8?"

Response: 4 and 2, and 8 and 1.

"Let's try 8 and 1 first," I say.

$$(8x - 3)(x - 1) = 8x^2 - 11x + 3\text{: } \textit{Nope.}$$

"What happens if I change the 1 and the 3 around?" Nope. That won't work because there will be an outer product of $-24x$ and inner product of $-x$.

"OK, let's try 4 and 2. Try it in your notebooks." I walk around and pick a student who gets it right to write it on the board.

$$(4x - 1)(2x - 3) = 8x^2 - 2x - 12x + 3 = 8x^2 - 14x + 3$$

I try another, a bit more involved, and work with them on it.

$$6x^2 - 25x + 14$$

I establish with the students that there will be two negative factors. Also, the factors for 14 are –14 and –1; and –7 and –2. Factors for 6 are 6 and 1 and 3 and 2.

$(6x - 14)(x - 1)$? *Nope! Middle term is*: $-6x - 14x = -20x$.

"Can you tell if we should switch the 14 and 1 around?" I want them to be able to see that if they do so they'll have $-84x$ and $-1x$, which do not sum to $-25x$.

Similarly, they should be able to see that –7 and –2 result in too big a middle term.

"Let's try $3x$ and $2x$ as our first terms."

$(3x - 1)(2x - 14)$ *Nope! Middle term is* $44x$.

"What should we try next?"

They will suggest –7 and –2. These work:

$$(3x - 2)(2x - 7)$$

Examples. I will work with them on the first and then they work independently.

1. $2a^2 + 9a + 9$. *Answer*: $(2a + 3)(a + 3)$.
2. $3x^2 - 10x + 8$. *Answer*: $(3x - 4)(x - 2)$.
3. $6x^2 + 7x - 3$. *Answer*: $(3x - 1)2x + 3)$.
4. $6x^2 + x - 15$. *Answer*: $(3x + 5)(2x - 3)$.
5. $4x^2 - 12x + 9$. *Answer*: $(2x - 3)(2x - 3)$ *or* $(2x - 3)^2$.

This last one can be done easily be recognizing that the trinomial is a perfect trinomial square.

Homework

For the first day, problems should be a mix of trinomials with leading coefficient equal to 1, trinomials with prime numbers for the leading coefficient and final number, and non-primes for both leading coefficient and final number. For the second day, the mix should include trinomials in which the numbers are non-prime.

Splitting the Middle Method

There is a method of factoring these types of trinomials by "splitting the middle." Taking Problem 3 above, the leading coefficient and final number are multiplied, obtaining −18. Next, find factors of −18 that sum to 7. Those would be 9 and −2. Now, the trinomial's middle term of 7x is rewritten as $9x - 2x$ so that the trinomial is now $6x^2 + 9x - 2x - 3$.

Factor by grouping: $3x(2x + 3) - (2x + 3) = (2x + 3)(3x - 1)$.

I present this on the next day and present it as an option for those who find it easier to work with this method. Some students like it; others prefer the method presented on the first day.

13. WORKING WITH FACTORS WHOSE PRODUCT IS ZERO

PRIOR CONTENT

Previous to this lesson, students will have learned to completely factor an expression. Thus, the expression $3x^2 - 27$ is first factored as $3(x^2 - 9)$, which then can be further factored as $3(x - 3)(x + 3)$. A common mistake students make is to omit the 3 at the front of the factored expression.

Complete factoring also addresses expressions like $-x^2 + 4x - 4$, in which -1 is factored:

$$-1(x^2 - 4x + 4) = -1(x - 2)^2$$

Complete factoring is especially important in making factoring easier as shown below:

$$5x^2 - 25xy - 250y^2 = 5(x^2 - 5xy - 50y^2) = 5(x - 10y)(x + 5y)$$

PRESENT LESSON

This lesson addresses the zero property of multiplication; that is, if $ab = 0$, and a does not equal zero, then b must equal zero—and vice versa. This property is used to solve quadratic equations that lend themselves to factoring. Later, students will learn to solve quadratic equations which are not factorable. It is important that students continue to practice solving factorable quadratics as preparation for the future chapter.

WARM-UPS

1. Factor completely. $4x^4 - 64y^4$.

   Answer: $4(x^4 - 16y^4) = 4(x^2 - 4y^2)(x^2 + 4y^2) =$
   $4(x - 2y)(x + 2y)(x^2 + 4y^2)$.
2. Write in factored form. $x^2 - 2xy - 35y^2$. *Answer*: $(x - 7y)(x + 5y)$.
3. Solve. $\frac{1}{4}x - \frac{1}{5}x = 3$. *Answer*: $LCD = 20$; $20(\frac{1}{4}x) - 20(\frac{1}{5}x) = 20 \cdot 3$;
   $5x - 4x = 60$ $x = 60$.
4. Solve for x. $3{,}560 \cdot 52 \cdot \pi \cdot x \cdot \sqrt{35} = 0$. *Answer*: $x = 0$.
5. Write with positive exponents. $\frac{2xy^{-10}}{x^5}$. *Answer*: $\frac{2}{x^4y} \cdot 2 = \frac{4}{x^4y}$.

Problem 1 requires factoring of each successive difference of squares. Problem 2 has two variables; students must identify the factors of $-35y^2$ as $-7y$ and $5y$, or $7y$ and $-5y$. Problem 3 can be solved by multiplying each term by a common denominator. Problem 4 requires students to see that x equals 0 divided by the product of the numbers that are multiplied by x. The problem serves as a segue to the day's lesson. Problem 5 reviews negative exponents and the zero exponent. In the case of the zero exponent, it is important for students to realize that $2g^0$ is $2(1)$ and is not the same as $(2g)^0$.

THE ZERO PROPERTY RULE

I point out that Problem 4 of the Warm-Ups illustrates the principle that anything times 0 is 0.

"For example, suppose we have $32 \cdot 5x = 0$. Since any number times 0 is 0, you can see that x must equal 0. This illustrates the zero property rule."

A product is zero if, and only if, at least one of the factors is zero.

"If we have an equation like $xy = 0$, the equation will be true if $x = 0$, or $y = 0$, or both. Suppose we have an equation like $5(x - 5) = 0$. What does the zero property rule tell us?"

If there are no immediate responses, I supply a hint. "We have two factors, one of which is 5 and the other is $(x - 5)$. Look at the zero property rule."

Generally, a few students will see the connection and tell me that $x - 5$ is 0.

"Right; we know that 5 is not 0, so that only leaves one other factor. And since we now know that $x - 5 = 0$, can we solve it?"

This is a simple equation. They've solved many like it; yet, for some students it suddenly seems unfamiliar. In the context of new information, even the familiar can seem strange. I offer a hint. "How do we isolate x?"

This usually does it and they solve it.

"How would you solve $(x - 1)(x - 4) = 0$ using the zero property?"

I ask them to look at the rule again if there are no immediate responses, but there usually are a few as students catch on. Someone will say that either one of the factors must equal zero.

"That means that either $(x - 1)$ is 0, or $(x - 4)$ is 0. What then are solutions for x that make this equation true?"

Response: $x = 1$ and 4.

"Yes. If x equals 1, when plugging that into the equation we get $0 \cdot (x - 4)$, which equals 0. Same deal with $x = 4$. We have two solutions for this equation. We call this the "solution set," which for this equation is 1, 4."

Examples

1. $13(a - 2) = 0$. *Answer*: $a = 2$.

Again, students should see that since 13 is not equal to 0, then $a - 2$ must equal 0. Alternately, both sides can be divided by 13 if that makes it more understandable for some students.

2. $-11(5 + c) = 0$. *Answer*: $c = -5$.
3. $37\left(16 - \frac{2}{y}\right) = 0$. *Answer*: $16 = \frac{2}{y}$; $16y = 2$; $y = \frac{2}{16}$ *or* $\frac{1}{8}$.

Students may want to write 16 as $\frac{16}{1}$; they can then cross-multiply.

4. $(x - 3)(x - 5) = 0$. *Answer*: $x = 3,5$.

Solving Polynomial Equations by Factoring

"This last equation is what is called a quadratic equation. These are equations that are in the following form." I write on the board.

$$ax^2 + bx + c = 0$$

"These have a squared term, which makes this an equation of degree 2, since 2 is the highest exponent of all the monomials. Let's try one." I have them write in their notebooks: $x^2 - x = 6$.

Step 1: Put equation into standard form.

"Standard form means all terms are on left side, and right side = 0. And the powers should be descending. So is $x^2 - x - 6 = 0$ in standard form?" General agreement ensues.

Step 2: Factor the left side of the equation.

I ask them to do this. $(x - 3)(x + 2) = 0$.

Step 3: Set each factor equal to zero.

"This is the zero product property; we've already solved a few like this. The factoring gives you two equations for you to solve: $x - 3 = 0$; $x + 2 = 0$."

Step 4: Solve the resulting equations.

I ask them to solve it. $x = 3$, $x = -2$.

Step 5: Check each root in the original equation.

"Let's do this. Does it work?" I ask them to check it.

"What if we have something like this?"

$$2x^2 + 10x - 28 = 0$$

"Can we do something to make this easier to factor?"

Response: Factor out the 2.

Doing so results in $2(x^2 + 5x - 14) = 0$.

"We have two factors. We can divide by 2 if you want, or you can just look at it, and using the zero product property, we know at least one of the factors *must* be zero. So, which one is *not* zero?" They quickly tell me it is 2.

"Now, factor the trinomial and tell me the solutions."

$$(x + 7)(x - 2) = 0; x = -7, x = 2$$

Examples

1. $x^2 + 9x = -14$. *Answer*: $x^2 + 9x + 14 = 0$; $(x + 7)(x + 2) = 0$; $x = -7$ *and* -2.
2. $w^2 - 16 = 0$. *Answer*: $(w + 4)(w - 4) = 0$; $w = -4$, $w = 4$.
3. $2v^2 + v = 0$. *Answer*: *Factor out v*: $v(2v + 1) = 0$; $v = 0$ $v = -\frac{1}{2}$.

Some students get confused by this. Seeing the v by itself does not always trigger that it must equal zero. I ask, "What if it were $2v = 0$?" They have no problem seeing that v equals 0. "So instead of $2v$ we have $v(2v + 1)$, and it equals 0. We then have two equations: $v = 0$; $2v + 1 = 0$." Seeing this does not come immediately to some students, so it is important to continue to give equations of this type; they eventually get it, and they usually remark that they don't know what they found so difficult.

4. $x^2 = 15 + 2x$. *Answer*: *Put in standard form*: $x^2 - 2x - 15 = 0$; $(x - 5)(x + 3) = 0$; $x = 5, -3$.
5. $x^2 + 6x + 9 = 0$. *Answer*: $(x + 3)^2 = 0$; $x = -3$.

There is only one solution to this. Some students will not recognize that it is a perfect trinomial square and will factor it as $(x + 3)(x + 3) = 0$.

6. $x^2 = 3x$. *Answer*: $x^2 - 3x = 0$; $x(x - 3) = 0$; $x = 0,3$.

This is similar to Problem 3 above, except that it isn't in standard form. "If you don't put it in standard form, you might decide to divide both sides by x and get $x = 3$, and think you're done. But you're not. You've forgotten a solution. In standard form you will get two solutions."

Homework

Problems will be like those in the examples; a mix of trinomials and other expressions that can be factored.

14. ALGEBRAIC FRACTIONS (RATIONAL EXPRESSIONS): SIMPLIFYING

After the unit on factoring, students move into algebraic fractions, or rational expressions. Students' work with fractions in lower grades is now extended to algebraic forms.

An initial lesson in fractions occurs before the one presented here. Students learn that an algebraic fraction or rational expression is defined as the quotient of two algebraic expressions and that such fractions are defined only when the denominator is not zero. A fraction like $\frac{3}{x}$ has meaning, therefore, only when x is not equal to zero, and this is often written as a restriction: $x \neq 0$. Zero is an "excluded value"; that is, a number excluded from the set of numbers that x can be. Finding excluded values proves to be difficult for students when the fractions are more complex; for example:

$$\frac{1}{x-2}, \frac{5}{x^2-4}$$

For the first fraction, it's obvious that x cannot equal 2, but it can also be solved by writing the equation $x - 2 = 0$, or to ensure that students understand what x cannot equal, writing it as $x - 2 \neq 0$. As mentioned previously, although students have solved such a simple equation before, in new contexts the familiar seems strange—even more so for the second fraction, which requires factoring:

$$x^2 - 4 \neq 0; (x-2)(x+2) \neq 0, x \neq 2, -2$$

To alleviate confusion I increase students' familiarity by including these simple equations as part of the daily Warm-Ups.

Simplifying algebraic fractions is done by cancelling; for example:

$$\frac{a\not{c}}{b\not{c}d}, \frac{a}{bd}$$

More complex fractions require factoring. For example, the following simplification requires knowledge of and facility with factoring:

$$\frac{x^2 - 16}{2x - 8} = \frac{\cancel{(x-4)}(x+4)}{2\cancel{(x-4)}} = \frac{x+4}{2}$$

To find the excluded value, the final simplified version is not what we look at to determine it. The original expression is what is used—once seen it cannot be unseen, and an examination shows that $2x - 8$ cannot equal 0. Therefore, the excluded value is $x \neq 4$.

Warm-Ups

1. A rectangle is 5 feet longer than it is wide. The area of the rectangle is 66 square feet. Find its dimensions. *Answer*: *Let* w = *width*; *then* $w + 5$ = *length*. $w(w+5)=66$; $w^2 + 5w = 66$; $w^2 + 5w - 66 = 0$; $(w + 11)(w - 6) = 0$; $w = 6$, $w + 5 = 11$.

   $(w + 11)=0$ *is not considered. Dimensions can only be positive values.*
2. Solve. $x^2 = 8x - 15$.

   Answer: $x^2 - 8x + 15 = 0$; $(x - 5)(x - 3) = 0$; $x = 5,3$.
3. Solve. $x^2 = 17x$. *Answer*: $x^2 - 17x = 0$; $x(x - 17) = 0$; $x = 0, 17$.
4. Simplify. $\frac{x^2y}{x}$. *Answer*: xy.
5. Find the excluded value. $\frac{x}{x-5}$. *Answer* $x \neq 5$.

Students have had word problems such as in Problem 1 in the unit on factoring. Rectangle area problems are based on length × width = area. Negative values are discarded as answers, since all lengths must be positive. Problems 2 and 3 require putting the equation in standard form first. For Problem 3, both x and $x - 17$ are equal to 0; $x = 0$ is a solution frequently overlooked. Students have had problems like Problem 4 in working with powers, and now as part of the working with fractions.

Simplifying Fractions by Factoring

I refer to Problem 4 of the Warm-Ups as an example of simplifying a fraction. "You've been doing these for a while. Here's one that's a little different. See if you can do it." I write on the board:

$$\frac{x+5}{x+5}$$

It doesn't take long for students to say it equals 1. "Correct. Any number divided by itself is equal to 1. Is there an excluded value?"

Response: −5.

"Good. Now, we treat $x + 5$ as a single number that can be divided just as if we had $\frac{5}{5}$. What if I had something like this? Could we simplify it?"

$$\frac{2x+4}{x+2}$$

If no response, I ask, "Can we factor the numerator?"

In fact the numerator can be factored into $2(x + 2)$, and I then ask if we can cancel anything. They see that in the resulting expression $\frac{2(x+2)}{(x+2)}$ the $(x + 2)$ expressions can be cancelled. Provided of course that $x \neq -2$.

"When you simplify a fraction, look to see if it can be factored; factor as much as you possibly can and then cancel. Here's one that's slightly more complex."

$$\frac{3x-6}{x^2-5x+6}$$

I check their notebooks as they work it and select a student who has it correct to write the steps and solution on the board:

$$\frac{3(x-2)}{(x-2)(x-3)} = \frac{3}{x-3}$$

"Can anyone tell me what the excluded value is?" I will hear x cannot equal 3.

"Anything else? Remember, we look at the original expression to identify excluded values."

Eventually I hear x also cannot equal 2, which is correct.

Examples: I ask for excluded values as well.

1. $\frac{2d^2-2}{d+1}$. *Answer:* $\frac{2(d^2-1)}{d+1} = \frac{2(d-1)(d+1)}{d+1} = 2(d-1)$; $d \neq -1$.

I am watchful for students who don't readily recognize the difference between two squares.

2. $\frac{5xy^2}{20xy}$. *Answer:* $\frac{y}{4}$ $x \neq 0, y \neq 0$.

I may need to remind students that even though the simplified fraction does not have x or y in the denominator, the original fraction did—so those values must be excluded.

3. $\frac{6-t}{t^2-36}$.

Answer: Students become confused when they factor the denominator and see that it doesn't match the numerator:

$$\frac{6-t}{(t-6)(t+6)}$$

Which allows me to move to the next topic.

FACTORING OUT –1

"Some time in the past when we were on the factoring unit, you may recall that for problems like $-x^2 + 5x - 6 = 0$ we did something. What was it?"

This jogs some memories as they recall that factoring out –1 changes the signs inside the parentheses. "So, what can we do with our problem to get the numerator to match one of the expressions in the denominator?" I have them work on mini-whiteboards to show me:

$$\frac{-1(t-6)}{(t-6)(t+6)} = \frac{-1}{t+6} = -\frac{1}{t+6}$$

Someone may ask if they can factor out –1 in the denominator, and the answer is yes, you can. In either case, the value of the resulting simplified fraction is negative. Excluded values are 6 and –6 .

I have them try one more: $\frac{1-a^2}{a-1}$.

$$\frac{-1(a^2-1)}{a-1} = \frac{-1(a-1)(a+1)}{a-1} = -(a+1) = -a-1$$

I ask for the excluded value which is a $\neq$ 1.

Illegal Cancelling

"There is one thing you cannot do, and I'm willing to bet \$100 that at least one of you will do this—and hopefully not on a test or quiz." I write out the offending operation:

$$\frac{3 + \cancel{x}}{\cancel{x}} = 3 + 1 = 4; \textit{ WRONG}$$

"The only time you can cancel terms is if the terms are multiplied. If we have $\frac{3x}{3}$ the 3's can be cancelled. But not if the 3 and x are being added or subtracted." I have had a poster to this effect on the wall and have even embedded it on tests or quizzes, and I still have one or two students do this illegal cancelling. Usually after getting problems marked wrong on a test, they get the idea.

"Since I'm talking about things that drive me nuts, I'm still seeing people writing $(x + y)^2$ as $x^2 + y^2$. That's wrong! Use the rule for squaring binomials, or multiply the binomials if you have to."

Homework

Homework should be problems simplifying fractions, which involve some factoring—including factoring out −1 to reverse terms. For at least 10 problems I will require that they identify excluded values.

15. ADDING AND SUBTRACTING FRACTIONS

Prior to this lesson, students have learned multiplication and division of algebraic fractions which they find straightforward. Examples of problems include:

$\frac{2}{3} \cdot \frac{3(x-4)}{4} = \frac{x-4}{2}$, $\frac{9}{x+2} \div \frac{36}{x^2+2x} = \frac{9}{x+2} \cdot \frac{x^2+2x}{36} = \frac{9}{x+2} \cdot \frac{x(x+2)}{36} = \frac{x}{4}$

Adding and subtracting algebraic fractions is more complex and is a prerequisite for understanding the procedure for solving equations with algebraic fractions. I generally use three lessons to go through it.

On the first day, I present fractions with like denominators. The next two days are devoted to fractions with unlike denominators. Addition and subtraction of algebraic fractions is an extension of students' knowledge of numerical fractions. Students are used to seeing denominators that are multiples (e.g., 3 and 6; 3 and 9; and 4, 8 and 16). Now they must extend the idea of multiples to letter representation. Finally, students must learn to factor denominators and see that $a + 5$ and $a^2 - 25$ are multiples (i.e., $(a + 5)(a - 5) = a^2 - 25$).

WARM-UPS

1. Tony and Dwight live 160 miles apart. At noon, each boy rides his bike toward the other. Dwight travels 20 mph faster than Tony. They meet at 4 p.m.. What is the average speed of each person?

   Answer: *Let* x = *Tony's speed*; *then* $x + 20$ = *Dwight's speed.*

   Time travelled is 4 *hours.*

   Tony's distance in 4 *hr* + *Dwight's distance in* 4 *hr* = 160.

$4x + 4(x + 20) = 160$; $8x + 80 = 160$; $8x = 80$; $x = 10$; $x + 20 = 30$.

2. Find the numerator for the second fraction: $\frac{3}{2m^2} = \frac{?}{8m^2n}$. *Answer*: $3 \cdot 4n = 12n$.
3. Simplify. $\frac{x^2-1}{x+1}$. Answer: $\frac{(x-1)(x+1)}{x+1} = x - 1$.
4. Add and simplify. $\frac{1}{2} + \frac{3}{2}$. *Answer* $\frac{4}{2} = 2$.
5. Add and simplify. $\frac{x}{2} + \frac{3x}{2}$ *Answer* $\frac{x+3x}{2} = \frac{4x}{2} = 2x$.

Students will likely need guidance for Problem 1; a diagram helps. In this problem, the time travelled by each person is given—4 hr (from noon to 4 p.m.). Problem 2 requires students to see that whatever the multiplier is to change the denominator to $8m^2n$, it must then be multiplied by the numerator of the first fraction. For Problem 3, students must see that the numerator is the difference of two squares, and then to factor it and cancel the $(x + 1)$ terms. Problem 4 is given as a scaffold for students to solve Problem 5. These two problems segue to the day's lesson.

FRACTIONS WITH LIKE DENOMINATORS

After going through the last two Warm-Ups, I point out what should now seem obvious.

"Adding and subtracting algebraic fractions is done the same way as with numeric fractions. And you've already done that for Problem 5 of the Warm-Ups. Here is the formal rule."

$$\frac{a}{b} + \frac{c}{b} = \frac{a+c}{b} \textit{ and } \frac{a}{b} - \frac{c}{b} = \frac{a-c}{b} \quad b \neq 0$$

I now use examples as a means for practice and instruction.

EXAMPLES

1. $\frac{4}{9} + \frac{7}{9} - \frac{y}{9}$. *Answer*: $\frac{4+7-y}{9} = \frac{11-y}{9}$.

I write the addition and subtraction step over the common denominator so students get used to doing it in this fashion. It helps them see what is actually going on and makes it clear when there is a subtraction of a binomial, the signs inside the parentheses must change, as is the case in the next problem.

2. $\frac{5}{p+q} + \frac{q}{p+q} - \frac{q-p}{p+q}$. Answer: $\frac{5+q-(q-p)}{p+q} = \frac{5+q-q+p}{p+q} = \frac{5+p}{p+q}$.

I work through this with the class so they see that the minus sign outside $(q - p)$ changes the signs inside. I also want students to see $(p + q)$ as a single entity.

3. $\frac{2x}{9x^2} + \frac{9}{9x^2}$. *Answer:* $\frac{2x+9}{9x^2}$.

Simplifying Sums and Differences

"Looking at the last example, let's say we had something slightly different."

$$\frac{3x}{9x^2} + \frac{9}{9x^2}$$

I have them do this on mini-whiteboards.

$$\frac{3x + 9}{9x^2}$$

"Can anything be factored? And can anything be cancelled?"

They will see that the 3 can be factored and cancelled:

$$\frac{3(x + 3)}{9x^2} = \frac{x + 3}{3x^2}$$

Examples

1. $\frac{4y}{3} - \frac{7y}{3} =$? *Answer:* $\frac{-3y}{3} = -y$.
2. $\frac{a}{a+b} + \frac{b}{a+b}$. *Answer:* $\frac{a+b}{a+b} = 1$.
3. $\frac{3a+2b}{4ab} + \frac{a+2b}{4ab} =$? *Answer:* $\frac{4a+4b}{4ab} = \frac{4(a+b)}{4ab} = \frac{a+b}{ab}$.
4. $\frac{7a}{a^2-b^2} - \frac{5a+4b}{a^2-b^2} + \frac{2b}{a^2-b^2}$. *Answer:* $\frac{7a-(5a+4b)+2b}{a^2-b^2} = \frac{7a-5a-4b+2b}{a^2-b^2} = \frac{2a-2b}{a^2-b^2} = \frac{2(a-b)}{(a-b)(a+b)} = \frac{2}{a+b}$.

I work with them on this last one to make sure they understand each of the steps and to show them to be careful when there is a subtraction.

5. $\frac{x^2}{x-3} - \frac{9}{x-3} =$? *Answer:* $\frac{x^2-9}{x-3} = \frac{(x-3)(x+3)}{x-3} = x + 3$.

Homework

Problems are a mixture of fractions that can be simplified and those that cannot. For some problems, I'll ask that excluded values be identified.

Day 2: Unlike Denominators

"In our last lesson, we worked with fractions that have like denominators. Today, we talk about unlike denominators. It involves finding the least common multiple between two numbers. What is the least common denominator for $\frac{1}{3}$ and $\frac{1}{9}$?" They answer it easily. "What about $\frac{1}{6}$ and $\frac{1}{9}$?" A slight pause, and they answer "18." "What about $\frac{1}{5}$ and $\frac{1}{7}$?" I'm told 35.

For this last, I point out that sometimes the least common multiple is the product of the two numbers.

"Algebraic multiples are similar. What is the common multiple of a and a^2?

Response: a.

"What about $6a$ and $9ab$?"

I discuss this so that students see how it works. "We first look at 6 and 9. What's the LCD?"

Response: 18.

"Then, we look at 18 and a and ab. What's the LCD?

Response: $18ab$.

"How about $5a$ and $7a$?"

Response: $35a$.

"What about $5a$ and $7x$?"

Some hesitation and prompting. The answer: $35ax$.

"Now. let's suppose we have these two fractions to add. We need to find the least common denominator."

$$\frac{a}{5} + \frac{b}{7x}$$

Step 1: Find the LCD:

"We know that the LCD is $35x$."

Step 2: Rename the two fractions with the new denominator.

I work through this one step at a time. $\frac{a}{5} \cdot \frac{7x}{7x} + \frac{b}{7x} \cdot \frac{5}{5}$. "Whatever we multiply by the denominator, we also have to multiply the numerator by the same."

Step 3: Find the sum or difference.

I ask the students to find the answer. $\frac{a}{5} \cdot \frac{7x}{7x} + \frac{b}{7x} \cdot \frac{5}{5} = \frac{7ax+5b}{35x}$.

"Now, you try with this one." $\frac{a}{6b} - \frac{1}{12b}$.

The LCD is $12b$. I ask a student to write the steps and the final answer on the board.

$$\frac{a}{6b} \cdot \frac{2}{2} - \frac{1}{12b} = \frac{2a-1}{12b}$$

Examples: For each one, I ask students to tell me the LCD first.

1. $\frac{4}{x^2} - \frac{3}{x}$. *Answer*: $LCD = x^2$; $\frac{4}{x^2} - \frac{3}{x} \cdot \frac{x}{x} = \frac{4-3x}{x^2}$.
2. $\frac{4b}{a} - \frac{1}{2}$. *Answer*: $LCD = 2a$; $\frac{4b}{a} \cdot \frac{2}{2} - \frac{1}{2} \cdot \frac{a}{a} = \frac{8b-a}{2a}$.
3. $\frac{1}{ab} + \frac{a-2}{bc}$. *Answer*: $LCD = abc$; $\frac{1}{ab} \cdot \frac{c}{c} + \frac{a-2}{bc} \cdot \frac{a}{a} = \frac{c+a(a-2)}{abc} = \frac{c+a^2-2a}{abc}$.
4. $\frac{4}{a^2b} - \frac{2-a}{ab^2}$. *Answer*: $LCD = a^2b^2$ $\frac{b}{b} \cdot \frac{4}{a^2b} - \frac{a}{a} \cdot \frac{2-a}{ab^2} = \frac{4b-a(2-a)}{LCD} = \frac{4b-2a+a^2}{a^2b^2}$.

In this problem, students need to be aware that they are multiplying $(2-a)$ by $-a$.

Homework

I work several of the more complex homework problems with the students as they do the homework in class. The homework is limited to the more straightforward problems, and is mixed with problems with fractions that have like denominators.

Day 3: More Complex Denominators

"Yesterday, you had problems where the denominators might be something like ab and $2a$. What's the common denominator?" They've seen this before and identify the common denominator as $2ab$.

I then ask student to tell me how to add the following fractions:

$$\frac{1}{ab} + \frac{1}{2a}$$

This is similar to the problems we did in the previous lesson.

$$\frac{1}{ab}\cdot\frac{2}{2}+\frac{1}{2a}\cdot\frac{b}{b}=\frac{2+b}{2ab}$$

"Now, suppose I have two denominators like this."

$$x^2-25;\ 2x+10$$

"What's the first thing you're going to do?" Someone might suggest that we factor, but if no one does, I do: "Do you think factoring might help?" This generally turns on a few lights.

$$(x-5)(x+5);\ 2(x+5)$$

"What would be the common denominator for what I have on the board?"

After allowing them to discuss it, the verdict on the common denominator is:

$$2(x-5)(x+5)$$

I now write on the board:

$$\frac{1}{x^2-25}=\frac{1}{2x+10}$$

I proceed to add these fractions by asking the students what to do as I write down the steps.

Step 1: Find the LCD by factoring each denominator.

We did that: LCD = $2(x+5)(x-5)$.

Step 2: Write the fractions with factored denominators.

$$\frac{x}{(x-5)(x+5)}-\frac{1}{2(x+5)}$$

"Take each factor the *greatest number of times* it appears in any denominator. In this case, each factor appears only once and they are 2, $(x+5)$ and $(x-5)$."

Step 3: Multiply each fraction by an equivalent fraction.

"What's missing from the first fraction's denominator?" 2. "How about the second denominator?" $(x - 5)$.

$$\frac{2}{2} \cdot \frac{x}{(x-5)(x+5)} - \frac{x-5}{x-5} \cdot \frac{1}{2(x+5)}$$

Step 4: Combine the fractions and simplify.

$$\frac{2x-(x-5)}{2(x-5)(x+5)} = \frac{2x-x+5}{2(x-5)(x+5)} = \frac{x+5}{2(x-5)(x+5)} = \frac{1}{2x-10}$$

"It's really the same procedure we did yesterday, except now we have to factor first. Let's try this one and see how we do."

$$\frac{5}{2a-2b} + \frac{1}{a^2-2ab+b^2}$$

Factoring first, we obtain: $\frac{5}{2(a-b)} + \frac{1}{(a-b)^2}$. Some students will recognize the second denominator as a perfect trinomial square; others will simply factor it, which is fine because that's the form I want:

$$\frac{5}{2(a-b)} + \frac{1}{(a-b)(a-b)}$$

"We take each factor the greatest number of times it appears in any denominator. In the first fraction, '2' appears once as does $(a - b)$, and what about the rest?"

Response: $(a - b)$ appears twice in the second fraction.

"Ah, so we have to use $(a - b)^2$ in the denominator. So what is the LCD?"

Response: $2(a - b)^2$.

To keep the momentum, I pick a student who I know has the answer to write the next step on the board:

$$\frac{5}{2(a-b)} \cdot \frac{a-b}{a-b} + \frac{1}{(a-b)(a-b)} \cdot \frac{2}{2}$$

I'll pick another student to combine and simplify:

$$\frac{5(a-b)+2}{2(a-b)^2} = \frac{5a-5b+2}{2(a-b)^2}$$

Someone might ask if we can cancel the 2's, to which I emphatically answer "No!"

I try one more before moving on to the next topic.

$$\frac{a}{a+b} - \frac{b}{a^2-b^2}$$

"Can we factor anything?"

Response: $a^2 - b^2 = (a - b)(a + b)$.

"So what would be the LCD?"

Response: $LCD = (a - b)(a + b)$.

Putting it all together we have: $\frac{a(a-b)}{(a-b)(a+b)} - \frac{b}{a^2-b^2} = \frac{a(a-b)-b}{a^2-b^2} = \frac{(a^2-ab-b)}{(a^2-b^2)}$

Factoring -1. "What if I had something like this?" I ask and write the following on the board:

$$\frac{1}{x-1} + \frac{x}{1-x}$$

Someone might say the LCD is $(x - 1)(1 - x)$ which is technically a way to do this. "We could do that," I say, but there's an easier way. At which point someone might remember that we can reverse the order of a binomial by factoring out −1.

"Let's do that."

$$\frac{1}{x-1} + \frac{x}{-1(x-1)}$$

"Now, we have –1 in the denominator. That makes the whole fraction negative.

$$\frac{1}{x-1} - \frac{x}{x-1}$$

"What do I get when I subtract?" Someone tells me, and I write it:

$$\frac{1-x}{x-1}$$

"What can I do to reverse either the numerator or the denominator?"

I'm told to factor out −1. "I'm going to factor it out of the numerator, though I could factor it out of the denominator. But let's see what we get."

$$\frac{1-x}{-1(1-x)}$$

"Can I cancel anything?"

I'm told we can cancel $1 - x$ leaving $\frac{1}{-1}$, which equals −1.

"Let's do another."

$$\frac{1}{x-1}-\frac{x}{1-x}$$

"After we factor out −1 in the denominator we have this."

$$\frac{1}{x-1}-\frac{x}{-(x-1)}$$

"I'm going to rewrite the second fraction. We know that since the denominator is negative, the whole fraction is negative." I write on the board:

$$\frac{1}{x-1}-\left(-\frac{x}{(x-1)}\right)$$

I do this rather than have the students come up with it, because it is not yet obvious to students that that's what needs to be done, and having students discuss it and come up with it themselves is an iffy proposition. It takes time and interferes with the flow that's going on.

"We're subtracting a negative. What happens when we subtract a negative?"

Since I've asked this question before in other contexts, they know the answer is "It becomes positive," or "We add."

"Good. So, how would I rewrite this?" They tell me:

$$\frac{1}{x-1}+\frac{x}{x-1}=\frac{1+x}{x-1}$$

Now that they've seen how such problems are handled, the rule that I now write on the board ties it all together:

When factoring out –1 in the denominator of a fraction, reverse the order of the terms, and change the sign of the fraction.

Examples. For the first two I work with the class, and then have them work independently.

1. $\frac{4x}{x+4}+\frac{2x}{x+3}$. *Answer*: *LCD* = (x + 4)(x + 3); $\frac{4x(x+3)+2x(x+4)}{(x+4)(x+3)}=\frac{4x^2+12x+2x^2+8x}{(x+4)(x+3)}=\frac{6x^2+20x}{(x+4)(x+3)}$.
2. $\frac{3a}{2a+6}-\frac{a-1}{a+3}$. *Answer*: *LCD*: 2(*a* + 3); $\frac{3a-2(a-1)}{2(a+3)}=\frac{a+2}{2(a+3)}$.

Students may ask whether the final answer should be in the form $\frac{a+2}{2a+6}$. I answer that it doesn't matter, but it's good to start with factored form first to see if anything can be cancelled.

3. $\frac{x+1}{x-y}-\frac{2}{y-x}$. *Answer*: $\frac{x+1}{x-y}+\frac{2}{x-y}=\frac{x+3}{x-y}$.

HOMEWORK

Homework should be a combination of fractions with unlike denominators with both simple and complex form, including those that require factoring out −1. Also, fractions with like denominators should be included. In particular, I like to ask students to solve the following:

$$\frac{x^2}{(x+1)^2}+\frac{2x+1}{(x+1)^2}$$

The answer simplifies to "1" if students see that the numerator is a perfect square trinomial:

$$\frac{x^2+2x+1}{(x+1)^2}+\frac{(x+1)^2}{(x+1)^2}=1$$

16. FRACTIONAL EQUATIONS

There are two types of fractional equations that a first-year algebra course covers. The first type, which was covered earlier in the course and also in the lesson prior to this (not included in this book), consists of equations that have fractional coefficients, such as:

$$\frac{3}{4}x + \frac{5}{6} = \frac{2}{3}x$$

The second type has variables in the denominator:

$$\frac{2}{3n} + \frac{1}{n} = \frac{5}{9}$$

Both types are solved in the same way: all terms are multiplied by a common denominator. In the second type, however, that denominator includes variables as well as numbers. Students have had experience with such denominators from their work with adding and subtracting algebraic fractions. When adding or subtracting fractions, however, the denominator is not eliminated, whereas with equations it is.

Some students will retain a common denominator; for example, taking the equation given above, they will treat it the same as if they are adding fractions:

$$\frac{2}{3n} + \frac{1}{n} = \frac{5}{9};\ LCD = 9n$$

$$\frac{2}{3n} \cdot \frac{3}{3} + \frac{1}{n} \cdot \frac{9}{9} = \frac{5}{9} \cdot \frac{n}{n} \rightarrow \frac{15}{9n} = \frac{5n}{9n}$$

I will then suggest they eliminate the denominator $9n$ by multiplying both sides by the same, resulting in $15 = 5n$. It is then a matter of showing that this same result can be obtained by multiplying each term by $9n$.

I devote two days for this topic. After presenting the method of multiplying through by a common denominator on the first day, I then show how certain problems lend themselves to cross-multiplication. For example:

$$\frac{5}{x} = \frac{x-3}{2};\ LCD = 2x$$

This can be solved by multiplying each side by $2x$, which is effectively the same as cross multiplying.

$$2 \cdot 5 = x(x-3);\ 10 = x^2 - 3x$$

The cross multiplication method can be used for problems that have more than one term on each side:

$$\frac{2}{3x} + \frac{1}{x} = \frac{5}{9}$$

In this example, the fractions on the left-hand side are combined to form $\frac{5}{3x}$, while the right-hand side remains $\frac{5}{9}$. Cross multiplication can then be used.

I give problems that lend themselves to cross-multiplication as well as those for which multiplication of each term by the LCD is a better way to go.

WARM-UPS

1. The sum of two numbers is 46. Five times the smaller number is 6 more than twice the larger. Find the numbers. (*Hint*: This is a "hiding in plain sight" problem. If the smaller number is x, what is the larger number in terms of x?)

   Answer: x = *smaller number*; $46 - x$ = *larger number. and* $2(46 - x)$ = *twice the larger number. Equation is* $5x = 6 + 2(46 - x)$

   $5x = 6 + 92 - 2x$; $7x = 98$; $x = 14$, $46 - x = 35$.

2. Combine. $\frac{12}{x+2} - \frac{4}{x-2}$. *Answer*: $LCD = (x + 2)(x - 2)$; $\frac{12(x-2)-4(x+2)}{(x+2)(x-2)} = \frac{12x-24-4x-8}{(x+2)(x-2)} = \frac{8x-32}{(x+2)(x-2)}$ *or* $\frac{8x-32}{x^2-2}$.
3. Solve. $\frac{8}{x} = 2$. *Answer*: $2x = 8$; $x = 4$.
4. Solve. $\frac{x}{3} - \frac{x}{5} = 4$. *Answer*: $LCD = 15$; $5x - 3x = 60$; $2x = 60$; $x = 30$.
5. Factor as a difference of squares. $(a - b)^2 - 25$.

   Answer: $(a - b - 5)(a - b + 5)$.

Problem 1 is a "hiding in plain sight" problem. If x represents one of the numbers, then $46 - x$ represents the other. Suggested prompt: "If 10 is one of the numbers, how do you find the other?" The student then extends the pattern to x being one of the numbers. Problem 2 is subtraction of fractions. Some students after obtaining $\frac{8(x-4)}{x^2-4}$ may then try to cancel the fours. This is not permitted; the fraction cannot be simplified. Problem 3 requires students to see that both sides are multiplied by the denominator. Suggested prompt: "What if the problem were $\frac{x}{8} = 2$? How do you eliminate the 8?" Problem 4 is the same as $\frac{1}{3}x - \frac{1}{5}x = 4$ and requires that all terms be multiplied by the LCD, which is 15. Problem 5 requires students to see $(a - b)$ as a single number. Suggested prompts: "What if it were $x^2 - 25$?" "What's the square root of x^2?" "What's the square root of $(a - b)^2$?"

Day 1:

Multiplying by LCD

I use Warm-Up Problem 4 as the intro to today's lesson.

"You learned about equations with fractional coefficients and Problem 4 is one of those type of problems. I notice some of you when solving this, kept the common denominator."

$$\frac{x}{3} - \frac{x}{5} = 4; \frac{5x}{15} - \frac{3x}{15} = \frac{60}{15}$$

"Then, what did we say we should do to eliminate the denominator?"

Response: Multiply both sides by 15.

"You can do it this way, but it's building in an extra step. You can just multiply each term by 15 to begin with. It's much simpler and will be important for today's lesson which is about solving equations with fractions where the denominator has variables in the denominator." I write on the board:

$$\frac{7}{x-3} = \frac{2}{x+2}$$

I ask them to write down the steps as we solve the problem.

Step 1: Find the LCD. In this case, it is $(x - 3)(x + 2)$.

Step 2: Multiply each term of the equation by the LCD and cancel denominators.

$$\frac{\cancel{(x-3)}(x+2)}{1} \cdot \frac{7}{\cancel{x-3}} = \frac{(x-3)\cancel{(x+2)}}{1} \cdot \frac{2}{\cancel{x+2}}$$

Step 3: Solve the equation.

$7(x + 2) = 2(x - 3)$; $7x + 14 = 2x - 6$; $5x = -20$; $x = -4$

Step 4: Check for excluded values.

The original equation is used to identify excluded values which are $x \neq 3$, -2. Since the solution to the equation is -4, it is neither of the excluded values and therefore stands.

Examples. These are part of the homework assignment, so we kill two birds at once by getting them going on the homework via working the first few examples.

1. $\frac{12}{x+2} = \frac{4}{x-2}$. *Answer*: LCD = $(x + 2)(x - 2)$; $\frac{(x-2)(x+2)}{1} \cdot \frac{12}{x+2} = \frac{(x-2)(x+2)}{1} \cdot \frac{4}{x-2}$; $12(x - 2) = 4(x + 2)$; $12x - 24 = 4x + 8$; $8x = 32$; $x = 4$.

   Exluded values: $x \neq 2, -2$.
2. $\frac{1}{x} - \frac{2}{x} = 6$. *Answer*: *LCD* = x; $1 - 2 = 6x$; $6x = -1$; $x = -\frac{1}{6}$ *Excluded value*; $x \neq 0$;
3. $\frac{3}{5x} - \frac{3}{4x} = \frac{1}{10}$. *Answer*: *LCD* = $20x$; $12 - 15 = 2x$; $-3 = 2x$; $x = -\frac{3}{2}$. *Excluded value*: $x \neq 0$.

4. $\frac{2}{3x} + \frac{1}{x} = \frac{5}{9}$. *Answer: LCD* = 9*x*; 6 + 9 = 5*x*; 5*x* = 15; *x* = 3. *Excluded value x* ≠ 0.

5. $\frac{5}{m+6} = \frac{m-6}{m}$. *Answer: LCD* = *m*(*m* + 6); $5m = m^2 - 36$; $m^2 - 5m - 36 = 0$; $(m - 9)(m + 4) = 0$, $m = 9, -4$. *Excluded values: x* ≠ 0, −6.

I work with the students on this one and remind them how to use factoring when solving a quadratic.

Homework

Problems consist of both types of fractional equations. The problems should be relatively straightforward and also include two or three addition/subtraction of fractions that are not equations. The reason is that students occasionally get combining fractions confused with equations and try to eliminate the denominator.

Day 2

I go over any homework problems that students found difficult so that students understand the process used in solving them.

Solving by Cross Multiplication. I start out with a problem that was in their homework preferably, or something similar to it.

$$\frac{5}{x} = \frac{1}{2}$$

"In solving this, you multiplied by the LCD, which is what?"

Response: 2*x*.

They solve it and obtain *x* = 10.

"Now when you did that, you multiplied both sides by 2*x*, and ended up with 10 = *x*. This problem, in fact looks a lot like the proportion problems you did in seventh grade. And when you did those problems you did something called cross multiplying. Which you were still doing with this problem, and didn't realize it. Because watch:"

$$\frac{2x}{1}\left(\frac{5}{x}\right) = \frac{2x}{1}\left(\frac{1}{2}\right);\ 2(5) = x(1)\ ;\ and\, x = 10$$

"For a lot of fractional equation problems you can just cross multiply. Let's look at another one."

$$\frac{3n+1}{6n} = \frac{5}{12}$$

"I want this half of the room to multiply both sides by the LCD and this half to cross multiply."

I do this, monitoring their notebooks and pick two students representative of each side to write the solved problem on the board. Side by side, we have the multiply by LCD version and the cross multiply version:

LCD Version: $\frac{12n}{1}\left(\frac{3n+1}{6n}\right) = \frac{12n}{1}\left(\frac{5}{12}\right) \rightarrow 6n + 2 = 5n;\ n = -2$ *and*

Cross Multiplication Version: $36n + 12 = 30n;\ 6n = -12;\ n = -2$.

"Which one was easier, or are they both about the same?"

Generally, they say it's about the same, though they note that the cross multiplication method resulted in larger numbers.

"Let's look at this one:

$$\frac{3}{4x} + \frac{1}{x} = \frac{7}{8}$$

"We could multiply all terms by $8x$. Or we could combine the fractions on the left-hand side. So do that; add $\frac{3}{4}x + \frac{1}{x}$ and keep the right-hand side the same. Show me what you get." The result:

$$\frac{7}{4x} = \frac{7}{8}$$

"Now, you can solve it by cross-multiplication."

They do so and get $x = 2$. Some students say they solved it by inspection, seeing that in the denominators 4 times 2 is 8.

Finding What's Missing: Cancelling in Advance

"I just want to show you things you can do that may make things easier. There are times when multiplying by the LCD is probably the easiest way to solve the equation. For example," I say writing the following on the board:

$$\frac{4}{3t-2}+\frac{7}{3t}=\frac{1}{t}=0$$

"What's the LCD?" Some students will get confused and not see that $3t$ is a multiple of t. The general consensus is:

$$(3)(t)(3t-2)$$

"You could multiply each term by this chain of factors. Or you could do it like we did when adding and subtracting fractions where you identify what's missing from the denominator. You're canceling in advance."

I work through the problem with them to show them what I mean.

"The denominator of the first term is $3t - 2$. If we were to multiply by the LCD of $3t(3t - 2)$, we would cancel $(3t - 2)$ and end up multiplying the numerator, 4, by $3t$. We can skip the cancel part. We know that the denominator is missing $3t$ to form the LCD, so we just multiply 4 by $3t$ and get rid of the denominator. What would we multiply the 7 by in the second term?"

Students discuss this a bit and decide it will be $3t - 2$. "And what about the third term?" The answer comes back as $3(3t - 2)$. "What we have then is this."

$$4(3t)+7(3t-2)-3(3t-2)(1)=0$$

"Multiply it out, combine terms, and solve. Do it in your notebooks." I walk around and check:

$$12t+21t-14-9t+6=0;\ 24t-8=0;\ 24t=8;\ t=\tfrac{1}{3}$$

I usually get the response, "Can we do it the long way on something like this?" to which I respond, "Of course. I'm just showing you various things you can do. Let's do some easier ones."

Examples. Again, the examples are taken from the homework. Students can elect to use whatever method they wish, and I assure them that I won't require them to do a shortcut on a test or quiz if they do not wish to do so.

1. $\frac{6}{z-1} = \frac{z}{2}$. *Answer: Using cross multiplication*: $12 = z^2 - z$; $z^2 - z - 12 = 0$; $(z - 4)(z + 3) = 0$, $z = 4$, $z = -3$. *Excluded value*: $z \neq 1$.
2. $\frac{4}{3n} - \frac{n+4}{6n} = 2$. *Answer: Cross multiplication method*: $\frac{8-(n+4)}{6n} = \frac{2}{1}$; $\frac{4-n}{6n} = \frac{2}{1}$; $4 - n = 12n$; $13n = 4$; $n = \frac{4}{13}$. *Excluded value*: $n \neq 0$.

Students will think they have the wrong answer because they are used to getting whole numbers, not to mention that the fraction does not look "friendly."

3. $a - \frac{2}{a-3} = \frac{a-1}{3-a}$. *Answer: Reverse the denominator on the right-hand side*: $a - \frac{2}{a-3} = -\frac{a-1}{a-3}$; $LCD = a - 3$; $a(a - 3) - 2 = -(a - 1)$

   $a^2 - 3a - 2 = -a + 1$; $a^2 - 2a - 3 = 0$; $(a - 3)(a + 1) = 0$; $a = -1$.

Important! The excluded value is $a \neq 3$. Therefore, 3 is not a solution; only −1 is a solution. I emphasize that this is why we identify excluded values—sometimes what we get when we solve an equation is not a solution.

4. $c - \frac{c}{1-c} = \frac{2-c}{c-1}$. *Answer*: $c - \frac{c}{1-c} = -\frac{2-c}{1-c}$; $c(1 - c) - c = -(2 - c)$; $c - c^2 - c = -2 + c$; $-c^2 = -2 + c$. *Move $-c^2$ to the right-hand side.*

   $c^2 + c - 2 = 0$; $(c + 2)(c - 1) = 0$; $c = -2, 1$. *Excluded value*: $c \neq 1$. *Therefore,1 is not a solution; only −2 is.*

Some students may solve the problem by not moving $-c^2$ to right-hand side, resulting in this equation: $-c^2 - c + 2 = 0$. Negative one should then be factored: $-1(c^2 + c - 2) = 0$. Divide both sides by −1, leaving $c^2 + c - 2 = 0$, which is then solved as shown.

HOMEWORK

Problems should include mostly those that can easily be solved by cross multiplication but should also include some that are more complex, including those that have one of the solutions excluded.

17. WORD PROBLEMS THAT USE ALGEBRAIC FRACTIONS

Now that students are working with algebraic fractions, they have a tool that expands the types of word problems that they now can solve. For this book, I am focusing mostly on distance/rate/time problems, but algebraic fractions allow students to also solve problems involving work and mixtures. Typical problems of these categories include the following:

Work: One bulldozer clears land twice as fast as another. Together, they clear a large tract in 1 ½ hours. How long would the larger bulldozer take?

Solution: Let x = time for larger bulldozer to clear the tract alone. Then, $\frac{1}{x}$ represents the fraction of land bulldozed in 1 hour by the larger bulldozer, and $\frac{1}{2x}$ is the fraction of land bulldozed in 1 hour by the smaller one. $\frac{3}{2}\left(\frac{1}{x}+\frac{1}{2x}\right)=1$.

Therefore, $\frac{1}{x}+\frac{1}{2x}=\frac{2}{3}$; $6+3=4x$; $x=\frac{9}{4}$ or $2\frac{1}{4}$ *hr*.

Mixtures: To reduce 16 mL of a 25% solution of sulfuric acid to a 10% solution, how much distilled water needs to be added?

Solution: Let x= amount of water to be added; 25% of 16 mL is 4 mL. The ratio of the amount of acid to total solution is $\frac{4}{16}$. The amount of acid stays the same, but the total volume changes. The water increases the volume to $16 + x$. Therefore, the new ratio with water equals 10%:

$\frac{4}{16+x}=0.1$; $4 = 1.6 + 0.1x$; $0.1x = 2.4$; $x = 24$ *mL*.

If the above problem called for increasing the concentration of sulfuric acid to 60%, it would be solved as follows:

Solution: Let x = amount of sulfuric acid added. The initial ratio is the same: $\frac{4}{16}$. This time, however, the amount of acid is increased by x, which also increases the volume by the same amount. Therefore, x is added to the numerator *and* denominator: $\frac{4}{16+x} = 0.6; 4 + x = 9.6 + 0.6x$

$0.4x = 5.6; x = 14$ mL.

In this section, I describe lessons for round-trip problems, and wind and current problems. Each lesson is presented on a separate day. Previously, distance/rate problems were set up on the basis of "distance = distance." Since distance = rate · time, or $D = rt$, we can express time as $t = \frac{D}{r}$. Both types of problems are based on "time = time."

WARM-UPS

1. Solve. $\frac{2}{3c} + 1 = \frac{1}{2c}$. *Answer*: $LCD = 6c; 4 + 6c = 3; 6c = -1; c = -\frac{1}{6}\ (c \neq 0)$.
2. Write an equation and solve. 48 is 32% of what number? *Answer*: $48 = 0.32x; x = \frac{48}{0.32} = 150$.
3. Simplify and identify excluded value. $\frac{6x+6y}{6x-6y}$. *Answer*: $\frac{6(x+y)}{6(x-y)} = \frac{x+y}{x-y}\ x \neq y$.
4. Simplify; identify the excluded value. $\frac{x^2-1}{x^2+x}$. *Answer*: $\frac{(x-1)(x+1)}{x(x+1)} = \frac{x-1}{x}\ x \neq 0, -1$.
5. At 12 noon, two river steamers are 120 miles part. They pass St. Louis at 6 p.m. headed in opposite directions. If the northbound boat steams at 9 mph, find the rate of the southbound boat.

   Answer: Let x = speed of southbound boat. Fill out the chart. We know they meet in 6 hours since they start at noon and meet at 6 p.m.

	Rate × Time = Distance		
Northbound	9	6	54
Southbound	x	6	$6x$

Distance of southbound boat (54) + distance of northbound boat (6x) = 120

$54 + 6x = 120; 6x = 66; x = 11$ *hr*.

Problems 1–4 are problems students have worked on in the unit. Problem 5 is a distance = distance type problem.

Day 1

Round-Trip Problems

I make the following announcement to the class. "So far in this class, you've been solving distance/rate problems that are in the form of distance = distance." Students look at me as if waiting for the other shoe to drop.

"Today, we'll be working on problems that are in the form of time = time. Before we get into that, someone tell me the formula we've been using for distance in these problems."

Response: distance = rate times time.

I write $D = rt$ on the board. "If I wanted to express this equation in terms of t for time, how would I write it?"

Most students know this: $t = \frac{D}{r}$.

"This is the formula for time: distance over rate. We can use this to solve problems involving round trips, which is what we'll be doing today." I then project the following problem on the board:

A troop of scouts hiked to the scout cabin at the rate of 2 mph. They rode back to headquarters at the rate of 18 mph. If the round trip took 10 hours, how far is headquarters from the cabin?

Ignoring comments that the scouts aren't moving very fast, I ask: "What is the total time for the trip?"

Response: 10 hours.

"What are we trying to find?"

Response: The distance from headquarters to the cabin.

"We don't know how long it takes to get to the cabin, nor how long it takes to get back. What we know is that the time to the cabin and the time to return totals 10 hr. So, let's write that."

Time from HQ to cabin + time from cabin to HW = 10

"Do we know the rates of speed going to the cabin and returning? What are they?"

Response: 2 mph going to the cabin and 18 mph returning.

"We know the formula for time. We have the rates but not the distance. How should we represent distance then?

Some students will say D; others will say x. I will choose D to make it more obvious what we're doing.

"How do I represent the time going to the cabin at 2 mph?" I ask students to write in their notebooks, and seeing that some have the answer, I write it on the board: $\frac{D}{2}$.

"And how do I represent the time returning to headquarters at 18 mph?" Just about everyone gets it now: $\frac{D}{18}$.

"Plug those into the sentence I just wrote and give me an equation." I search for a correct equation; there are typically many: $\frac{D}{2} + \frac{D}{18} = 10$.

Solve it: $9D + D = 180$; $10D = 180$; $D = 18$ miles.

"Plug 18 into the original equation," I say. And sure enough: $\frac{18}{2} + \frac{18}{18} = 10$.

Examples. I work with the class on the first and perhaps second problems; the exercises are taken from the homework assignment.

1. Dick's motorboat averages 8 mph. On a trip, the motor breaks down. He rows back at 2 mph. When he returns to the dock, he has been gone 5 hours. How far has he rowed? *Answer*: $\frac{D}{8} + \frac{D}{2} = 5$; $D + 4D = 5$; $5D = 5$; $D = 1$ *mile*.
2. Amy walks to the repair shop at 2 mph, picks up her bike and rides home at 10 mph. If the round trip takes $1\frac{1}{2}$ hours, how far is the shop from Amy's house?

   Answer: $\frac{D}{2} + \frac{D}{10} = \frac{3}{2}$; $5D + D = 15$; $6D = 15$; $D = \frac{15}{6} = \frac{5}{2}$ *or* $2\frac{1}{2}$ *miles.*

For this problem, I advise students to use $\frac{3}{2}$ for total time.

3. A man travels to a convention at an average speed of 50 mph. He has to return home immediately and flies back at an average speed of 300 mph. Total traveling time is $1\frac{3}{4}$ hours. How long did it take for him to fly back? What was the distance from home to the convention? Answer: $\frac{D}{50} + \frac{D}{300} = \frac{7}{4}$; $LCD = 300$; $6D + D = 75 \cdot 7 = 525$; $7D = 525$; $D = 75$ *miles. Time to fly back is* $\frac{D}{300} = \frac{75}{300} = \frac{1}{4}$ *hr or* 15 *min.*

Homework

The problems are a combination of round-trip problems as well as what students have solved before: opposite direction and same direction (catch-up) problems.

Day 2

Wind and Current Problems

After going through Warm-Ups and any homework problems that students want to see worked out, I project the following problem on the board.

> *Jim rows 9 miles downstream in the same time that he rows 3 miles upstream. The current flows at 6 mph. How fast does Jim row in still water?*

The students have a deer-in-the-headlights type look as they read it. "I know what you're thinking," I say. "You're thinking there's no way you can do this problem."

Response: Correct!

"I don't believe you," I say. "And here's why. You recall we did some round-trip problems some time ago—like yesterday. They were based on 'time = time.' So is this one." I ask someone to read the first line of the problem.

"There are two distances stated here. One is a downstream distance, and the other is an upstream distance. What are the distances?"

Response: 9 miles downstream and 3 miles upstream.

"Downstream means going with the current. The water is flowing at what speed? Look at the problem."

Response: 6 mph.

"If there were no current at all, do we know his speed in the water?" There is general agreement that we do not.

"Let's let x be his speed with no current. How would I represent his speed with a 6 mph current?"

Some students will mistakenly say $x - 6$. If no other students intercede, I will, asking if rowing with the current results in a faster or slower speed. Seeing the light now, they say the speed is $x + 6$. They also see that his upstream speed would be slower: $x - 6$.

"Let's put all this information in a table," I say, and draw a table underneath the problem. I ask students to copy this in their notebooks.

	Distance	÷ *Rate*	= *time* = $\frac{D}{r}$
Downstream			
Upstream			

"The time to row 9 miles downstream is *the same as* 3 miles downstream. What does 'the same as' mean?"

Response: Equal.

"How do we represent time?"

Response: Distance/rate.

We now fill in the table.

	Distance	÷ *Rate*	= *time* = $\frac{D}{r}$
Downstream	9	$x + 6$	$\frac{9}{x+6}$
Upstream	3	$x - 6$	$\frac{3}{x-6}$

"Now we write 'time downstream = time upstream' as an equation." I ask students to write it in their notebooks and solve.

$$\frac{9}{x+6} = \frac{3}{x-6};\ 9x - 54 = 3x + 18;\ 6x = 72;\ x = 12\ mph$$

To check it, we plug 12 into the original equation. $\frac{9}{18} = \frac{3}{6}$; *yes*, $\frac{1}{2} = \frac{1}{2}$.

Examples. I use the homework problems as examples, and I work through the first two with students. Then, I provide guidance as they work on the rest.

1. The riverboat Memphis Belle sailing at the rate of 18 mph in still water can go 63 miles down the river in the same time it takes it to go 45 miles up the river. What is the speed of the current in the river?

   Answer: Let c = speed of current; $\frac{16}{18+c} = \frac{45}{18-c}$.

After we get to this point, I ask, "Can 63 and 45 be divided by anything?" They tell me 9 goes into both.

"We can then divide both numerators by 9 first, to make the numbers easier to work with." Students ask me if that's legal.

"It's the same thing as multiplying both sides by $\frac{1}{9}$." We go ahead and do it.

$$\frac{7}{18+c} = \frac{5}{18-c};\ 7(18-c) = 5(18+c);\ 126 - 7c = 90 + 5c;\ 12c = 36;\ c = 3$$

2. In still water, Jim's motorboat is 4 times as fast as the current in Pony River. He takes a 15-mile trip up the river and returns in 4 hours. Find the rate of the current.

This is actually a round-trip problem: upstream time + downstream time = 4 hr.

Answer: Let c = speed of current; then Jim's motorboat is $4c$.

Downstream speed is $4c + c$ *or* $5c$. *Upstream speed is* $4c - c = 3c$.

$\frac{15}{3c} + \frac{15}{5c} = 4;\ \frac{5}{c} + \frac{3}{c} = 4;\ \frac{8}{c} = 4;\ 4c = 8;\ c = 2$ *mph.*

Homework

Problems are similar to the previous examples, with some variation. I work with students as they do the problems. I assign five or six problems. Usually students finish them all within the class period.

18. GRAPHING LINEAR EQUATIONS: POINT-SLOPE FORM

PRIOR CONTENT

After working with algebraic fractions, students move to graphing of linear equations. Much of this material was presented initially in seventh grade: representing linear functions, constant rate of change, slopes, and slope-intercept form of equations (see Part II.B, Chapters 22, 23, and 24). I provide a brief review of basics, emphasizing the equation of a straight line ($y = mx + b$) and the formula for finding the slope of a line:

$$\textit{Given two points } (x_1, y_1) \textit{ and } (x_2, y_2),\ m = \frac{y_2 - y_1}{x_2 - x_1}$$

Students graph a linear equation by finding the y-intercept and plotting additional points based on the slope. They also learn to find the equation of a line by identifying the y-intercept and determining slope by inspection.

POINT-SLOPE FORM OF LINEAR EQUATIONS

The point-slope form of linear equations allows for identification of the equation of a line given 1) its slope and a point through which the line passes, or 2) two points through which the line passes.

The formula for the point-slope form tends to be confusing at first because of the notation. An alternative method is available, but I believe the point-slope formula provides a better connection with what slope is.

For information purposes, I present the alternative method for those who may prefer it. For example, a line has a slope of 2 and passes through the point (3, 4). The coordinates and the slope provide information that can be plugged into the $y = mx + b$ equation; namely $x = 3$, $y = 4$, and $m = 2$.

The resulting equation is $4 = 2(3) + b$, or $4 = 6 + b$. Solving for b, we obtain $b = -2$, so the final equation is $y = 2x - 2$.

WARM-UPS

1. Write with positive exponents. $\frac{x^{-4}}{y^{-2}}$. *Answer*: $\frac{y^2}{x^4}$.
2. Kyle hiked up a hill at 3 km/hr and back down at 6 km/hr. His total hiking time was 3 hours. What was the distance up the hill, and how long did the trip up the hill take him?

   Answer: Table is provided in the Warm-Ups for students to fill in.

	Distance	÷ *Rate*	= *Time* = $\frac{D}{r}$
Up	D	3	$\frac{D}{4}$
Down	D	6	$\frac{D}{6}$

Time uphill + time downhill = 3.

$$\frac{D}{3} + \frac{D}{3} = 3;\ LCD = 6;\ 3D = 18;\ D = 6\ km;\ Time\ uphill = \frac{6}{3} = 2\ hrs.$$

3. Add. $\frac{2}{x+1} + 4$. *Answer*: $LCD = x + 1$; $\frac{2+4(x+1)}{x+1} = \frac{3+4x}{x+1}$.
4. Find the slope between the two points. (3,4),(5,−6).

   Answer: $\frac{-6-4}{5-3} = \frac{-10}{2} = -5$.
5. Solve. $2r^2 + 11r + 12 = 0$. *Answer*: $(r + 4)(2r + 3) = 0$; $r = -4$,

   $2r = -3$; $r = -\frac{3}{2}$.

Problem 1 can be solved by seeing that the term in the numerator gets pushed down to the denominator, and the exponent changed to positive 4. Similarly, the term in the denominator gets pushed up to the numerator, changing the sign of the exponent. Problem 2 is a time = time type equation, which was covered in the previous lesson. Problem 4 requires application of the slope. Problem 5 offers a review of solving a quadratic equation by factoring.

Initial Activity

To prepare, I make up a set of index cards with an ordered pair of numbers on each side. All numbers are derived from the same equation (e.g., $y = 4x - 2$). Ordered pairs might be $(0,-2)$, $(1,2)$, $(-2,-10)$ and so on. I pass out the cards so each student has one.

"You all have a card with an ordered pair on it. Find a partner, and between the two of you find the slope of your two ordered pairs."

It doesn't take long before students figure out that everyone obtained the same slope: 4.

"Wow, how do you think that happened?" I say. "Let's try it again. Find a different partner and figure out the slope."

The same thing happens, of course. One time, a girl, totally mystified, asked me, "How did you do that?" as if it were a magic trick.

"What do you think is happening here?" At least two students, but usually more, venture that the ordered pairs come from the same line.

"Yes; no matter what two points you pick on a straight line, you will always get the same slope. This is the basis for what is called the "point-slope form of a linear equation."

Point-Slope Form Using Numbers

"You figured out that the ordered pairs on the cards were taken from the same equation. And in fact, that equation was $y = 4x - 2$. The fact that the slope is the same for a straight line for any two points on that line allows us to solve some problems that you would think are impossible."

I write on the board:

$$m = 4, \textit{ line passes through } (2,6)$$

"I've written the slope of a line, and a point that the line passes through. I used the same equation as before: $y = 4x - 2$. Let's pretend we don't know this. The equation tells us the slope, which we saw was 4.

"And since it's a straight line, no matter what two points I pick, the slope formula will be 4. We know that the point, (2, 6) is on the line. Now I'm going to represent any other point on the line by the ordered pair (x, y). Since I have two points, I can use the slope formula to find the slope of the line, like this."

$$\frac{y-6}{x-2} = 4$$

"How do I eliminate the $x - 2$ from the denominator?" I might have to write the 4 on the right-hand side as $\frac{4}{1}$ to get them to see the process via cross-multiplication.

$$y - 6 = 4(x - 2)$$

Solving the equation for y, they obtain $y = 4x - 2$.

"Let's try another one. This line passes through (3, 1) and has a slope of 2." I ask students to follow in their notebooks.

"Like we did before, we'll pick any other point on this line and call it (x,y). Somebody tell me how we show the slope of a line between a point (x,y) and (3,1)."

Response: $\frac{y-1}{x-3} = 2$.

Solving it, they obtain $y - 1 = 2(x - 3)$; $y = 2x - 5$.

I may have them do one more just to get into the rhythm. Then, we're ready for the actual derivation of the formula.

Derivation of Point-Slope Formula

"We're ready to make a formula that you can use. To review: No matter what two points we pick on a straight line, what do we know about the slope?"

Response: It's always the same.

"Instead of using numbers this time, let's say our slope is m. The line passes through the point (x_1,y_1)," I say, drawing a line on a grid and labeling a point with those coordinates.

"We solve for y in the same way we just did when we were using numbers. We get the following."

$$\frac{y - y_1}{x - x_1} = m$$

"We cross-multiply like we did before, and we get this." $y - y_1 = m(x - x_1)$.

"This is the point-slope formula. Write it down, learn it by heart, or if you don't like to memorize, have it tattooed on your arm so you'll always have it handy. That way, you won't have to Google it. Now, let's put it to use."

Examples. I work through the first example with them.

1. Line passes through (–6,1) and $m = 3$.

I ask them to write down the steps.

$$y - y_1 = m(x - x_1)$$

Step 1: $(x_1, y_1) = (-6,1)$ So $x_1 = -6$, $y_1 = 1$.

"The point it passes through will be the ordered pair (x_1, y_1)." I ask students to write x_1 above –6, and y_1 above 1 to help them identify what to plug in to the formula.

Step 2: Substitute values into the formula: $y - 1 = 3(x - (-6))$.

Step 3: Solve for y. $y - 1 = 3x + 6$; $y = 3x + 7$.

2. Point: (1,3); $m = 5$. *Answer*: $x_1 = 1$, $y_1 = 3$; $y - 3 = 5(x - 1)$; $y = 5x - 2$.
3. Point: (7,1); $m = 1$. *Answer*: $y - 1 = x - 7$; $y = x - 6$.

Writing an Equation with Two Points and No Slope

"Now, let's say you are asked to find the equation of the line that passes through (–2,1) and (3,4). Can you do it?"

Someone might ask, "What's the slope?"

"Ah ha! No slope is given. How can we find the slope?"

There are usually more than a few students who know: Use the slope formula.

Examples. I work the first two with the class.

1. Line passes through (–2,1), (3,–4).

Step 1: Find the slope (–2,1); (3,–4); slope = $\frac{1+4}{-2-3} = \frac{5}{-5} = -1$.

Step 2: Proceed using same steps as before, picking either point as the (x_1,y_1) point.

Either point can be used for the (x_1,y_1) point. I pick (3,–4) as the (x_1,y_1) pair and have other students use (–2,1).

(3,–4): $y-(-4) = -1(x-3) \rightarrow y+4 = -x+3;\ y = -x-1$.

(–2,1): $y-1 = -1(x+2) \rightarrow y-1 = -x-2;\ y = -x-1$.

2. Line passes through (–5,–5) and (–3,3). *Answer*: $m = \frac{8}{2} = 4$;

$y-3 = 4(x+3);\ y-3 = 4x+12;\ y = 4x+15$ or

$y-(-5) = 4(x-(-\!-5));\ y+5 = 4x+20;\ y = 4x+15$.

3. Line passes through (–2,–5); $m = \frac{7}{2}$.

Answer: $y+5 = \frac{7}{2}(x+2);\ y+5 = \frac{7}{2}x+7;\ y = \frac{7}{2}x+2$.

HOMEWORK

Problems are a combination of one point and slope given, or two points given.

Subsequent lesson is not included in this book, but introduces concept of parallel lines having same slopes, and perpendicular lines having slopes such that the slope of one line is the negative reciprocal of the slope of the other. Typical problems:

1. Line A is parallel to the line $y = 3x - 2$ and passes through the point (5,4). Find the equation of the line.

   Answer: $m = 3;\ y-4 = 2(x-5);\ y = 2x-6$.

2. Line A is perpendicular to the line $y = -2x + 6$ and passes through (0,3). Find the equation of the line. *Answer*: $m = \frac{1}{2};\ y-3 = \frac{1}{2}x;$ $y = \frac{1}{2}x+3$.

19. SYSTEMS OF EQUATIONS: MIXTURE PROBLEMS

Students have learned the basics of solving systems of equations in accelerated seventh-grade math or in Math 8 (see Part II.B, Chapters 25 and 26), Therefore, this section does not go into the methods for solving systems of equations. Typically, in an algebra 1 course, I offer a brief review since students remember the techniques fairly well. They are ready to solve more complicated problems with two variables, such as:

$$\frac{c}{5} + \frac{d}{2} = 1; \frac{d}{4} - \frac{c}{3} = -\frac{1}{2}$$

Students are introduced to solving word problems using two variables. In introducing two-variable word problems, I combine it with the use of the substitution method, which continues to pose difficulties for some students. I revisit the following problem:

> "The sum of two numbers is 46. Five times the smaller number is 6 more than twice the larger. Find the numbers."

This is what I call a "hiding in plain sight" problem. I start by reviewing how they have learned to solve it with one variable: Let x = the smaller number; then $46 - x$ is the larger number. (That's the part students continually find difficult). Then, the equation is $5x = 6 + 2(46 - x)$.

Using two variables, we let x = the smaller number and y = the larger number. The equations then become: $x + y = 46$ and $5x = 6 + 2y$. I suggest solving for y in the first equation, obtaining $y = 46 - x$.

We then substitute $46 - x$ in for y in the second equation:

$$5x = 6 + 2(46 - x)$$

I point out that this is the same equation using one variable and that they have actually been using the substitution method for some time, but didn't know it.[2] There is a collective "ah ha" from the class.

This section is focused on mixture problems. Up until this point, mixture problems have focused on how much water or other diluting material is needed to reduce a chemical mixture of a given percent, or how much pure chemical must be added to increase the percentage (see Part II Chapter 17.) These are most easily solved using one variable. Two-variable mixture problems involve combining two types of chemical mixtures of differing concentrations to make a mixture with a specific concentration.

Although not addressed in this book, I also teach word problems that focus on digits in numbers, age problems, fraction problems and cost problems. Examples of these follow:

Digit problem: "The sum of the digits of a two-digit numeral is 6. The number with the digits in reverse order is 12 times the original units digit. Find the original number. *Answer*: *Let* t = *tens digit and* u = *units digit*; $t + u = 6$; $10u + t = 12u$; $u = 2, t = 4$.

Age problem: "Ruth's father is 7 times as old as Ruth is. One year ago, he was 9 times as old as Ruth was. Find Ruth's present age." *Answer*: *Let* x = *Ruth's age*, y = *Ruth's father's age*. $y = 7x$; $y - 1 = 9(x - 1)$; $x = 4, y = 28$.

Fraction problem: "A fraction has a value of $\frac{3}{4}$. When 7 is added to its numerator, the resulting fraction equals the reciprocal of the original fraction. Find the original fraction." *Answer*: *Let* x = *numerator of the fraction*; y = *denominator of the fraction*; $\frac{x}{y} = \frac{3}{4}$; $\frac{x+7}{y} = \frac{4}{3}$; $x = 9, y = 12$.

Cost Problem: "Three chairs and a table cost \$149. Four chairs and a table cost \$177. How much does each item cost?" *Answer*: *Let* x = *cost of chair*, y = *cost of table*. $3x + y = 149$, $4x + y = 177$. $x = \$28, y = \$65$.

2 When I first started teaching algebra, I tried to teach the two variable method at the beginning of the year, so students would not get confused over the "hiding in plain sight type of problem." I found that it didn't work as well as I thought it would. I therefore teach it later, following the sequence in Dolciani et al.'s (1962) algebra book and others.

Warm-Ups

1. Two adult tickets and 3 student tickets cost \$12. One adult ticket and 2 student tickets cost \$7. How much does each kind of ticket cost?

   Answer: Let x = cost of adult ticket and y = cost of student ticket. $2x + 3y = 12$; $x + 2y = 7$; $x = 3, y = 2$.

2. Martin gets 27 questions right on a test. This is 90% of the total number of questions on the test. How many questions are on the test? (Write an equation and solve.) *Answer*: $0.9x = 27$; $x = 30$.

3. How much water must be added to 80 mL of a 60% iodine solution in order to dilute it to a 40% iodine solution? *Answer: Let x = amount of water; 60% of 80 mL = 48 mL, so* $\frac{48}{80+x} = 0.4$;

   $48 = 0.4(80 + x)$; $48 = 32 + 0.4x$; $0.4x = 16$; $x = 40$ *mL.*

4. Solve. $\frac{1}{a} + \frac{3}{2a} = \frac{1}{6}$. *Answer: LCD* $= 6a$: $6 + 9 = a$; $a = 15$.

   Cross multiplication method: LCD on left side: $2a$; $\frac{5}{2a} = \frac{1}{6}$; $2a = 30$; $a = 15$.

5. John spent \$32 and had \$48 left. What percentage of his money did he spend? *Answer: Total amount of money* $= x$; $x - 32 = 48$; $x = 80$; $\frac{32}{80} = \frac{2}{5} = 40\%$.

Students have had some problems similar to Problem 1 shown in class. Prompts: "Do we know the cost of each type of ticket?" "How do we represent the cost of 2 adult tickets?" Problem 2 is a percent question that students have had previously. Problem 3 is a mixture problem in which pure water is added to dilute the solution. Students have had such problems previously. Problem 5 requires students to find the total amount of money and write a proportion equation representing percent.

Conservation of Matter

I start out by writing the following on the board:

20 L of 25% alcohol solution + 10 L of 50% alcohol solution.

"I have two solutions of alcohol and water. How much alcohol is in the first solution?"

Response: 5 liters.

"And how much alcohol in the second solution?"

Response: 5 liters.

"And if I mix 20 L of solution with 10 L of solution, how much total solution do I have?"

Response: 30 liters.

"And in this 30 liter solution, how much alcohol is there?"

Silence.

"You just told me how much alcohol was in each solution." This usually gets them going, and I soon hear "10 liters."

"Correct." I pass out the illustration in Figure 19-1 for them to glue in their notebooks.

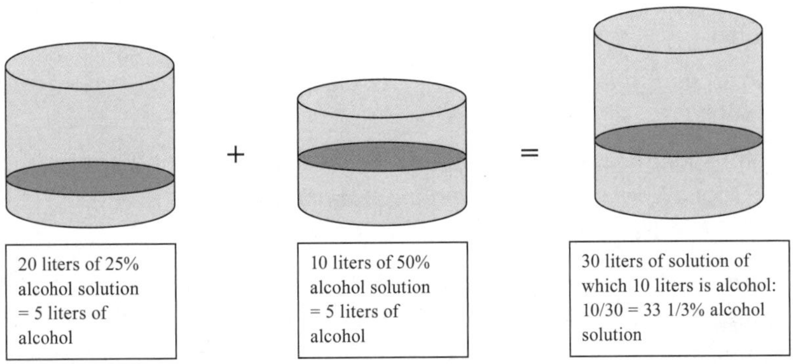
+
=
20 liters of 25% alcohol solution = 5 liters of alcohol
10 liters of 50% alcohol solution = 5 liters of alcohol
30 liters of solution of which 10 liters is alcohol: 10/30 = 33 1/3% alcohol solution

Figure 19-1

"The diagram shows what we just discussed, but there's some new information. It shows the percent of alcohol in the new mixture: $33\frac{1}{3}\%$. The percent of alcohol in the new mixture is different than the percentages in the two solutions. But the *amount* of alcohol in the two solutions is the same amount in the 30 liter solution—that is, 10 liters on the left-hand side of the equation, and 10 liters on the right. Which leads to this statement for you to write in your notebooks."

Material on left-hand side of the equation = material on the right-hand side of the equation.

"This is our prime directive, if you will, for the mixture problems we will be doing today."

Examples. I use examples to teach how to solve these equations, and when they're comfortable with it, I ask students to work independently, with guidance from me as needed.

1. A 35% sulfuric acid solution is mixed with a 65% sulfuric acid solution to obtain 20 ounces of a 41% solution. How many ounces of each solution are mixed?

"We're going to write two equations. The first equation is the 'amount of solution' equation. The second equation is 'amount of substance' equation. How much total solution is produced?

Response (from a few): 20 ounces.

"We'll let x = amount of 35% solution, and y= amount of 65% solution. That is, we're going to mix x ounces of the 35% solution with y ounces of the 65% solution, to get 20 ounces total. So what will the first equation be?" I may have to prompt a little more, but I'm soon given the correct answer: $x + y = 20$.

"This is the 'how much solution' equation. Now let's work on the 'amount of substance' equation. How do I represent the amount of sulfuric acid in the first solution? What is our percent?"

Response: 35%.

"So, we use 35% as a decimal, or 0.35. How do I write 0.35% of x ounces of solution?

Response: $0.35x$.

"Good, and now 65% of y ounces is written how?"

Response: $0.65y$.

"And now the final part. We want the total amount of the substance—in this problem it's sulfuric acid. What percent of sulfuric acid is in the total mixture? Look at the problem."

Response: 41%.

"And how many total ounces are there?"

Response: 20 ounces.

"Someone tell me the amount of sulfuric acid in that total mixture. How do you do it?"

I will usually hear the hesitant "I'm not sure, but…" and then 41% of 20, which is 8.2.

"Yes, 8.2 is correct, so let's write our second equation." $0.35x + 0.65y = 8.2$.

"And how do I get rid of the decimals?"

Response: Multiply everything by 100.

We do so, and I list the two equations:

$$x + y = 20; 35x + 65y = 820$$

"I strongly suggest you use substitution, letting $x = 20 - y$, but you can use elimination if that's easier for you. Go ahead and solve it."

They get $x = 16$ and $y = 4$.

2. How many pounds of a 70% copper alloy must be melted with how many pounds of a 90% copper alloy to obtain 100 pounds of an 81% copper alloy?

I lead them through this and check notebooks as they work the problem. *Answer*: *Let* x = *lb of* 70% *alloy and* y = *lb of* 90% *alloy.* $x + y = 100$; $0.7x + 0.9 y = 81$, *or* $7x + 9y = 810$. *Solve first equation for* x:$x = 100 - y$; *subsitute in second equation*: $7(100 - y) + 9y = 810$; $700 - 7y + 9y = 810$; $2y = 110$; $y = 55$ *lbs*, $x = 45$ *lb*.

3. A 60% iodine solution is mixed with a 30% iodine solution to produce a 50% solution of 10 mL iodine. How much of each solution are mixed?

Answer: *Let* x = *mL of* 60% *solution, and* y = *ml of* 30% *solution.*

Equation 1: $x + y = 10$. *Equation* 2: $0.6x + 0.3y = 5$;

$6x + 3y = 50$. *Elimination can be used*; *first equation is multiplied by* -3 *to obtain*: $-3x - 3y = -30$. $x = 6\frac{2}{3}$ *mL*, $y = 3\frac{1}{3}$ *mL*.

Homework

Problems should be mixed so that some can be solved with one variable, such as the dilution or addition of pure substance problem as discussed in Part III Chapter 17.

I also include a challenge problem using one variable, such as: “How many ounces of a 75% acid solution must be added to 30 ounces of a 15% acid solution to produce a 50% acid solution?” In this case, the equation would be: $0.75x + 4.5 = 0.5(x + 30)$.

$0.75x + 4.5 = 0.5x + 15$; $0.25x = 10.5$; $x = 42$ oz.

20. TWO-VARIABLE WIND AND CURRENT PROBLEMS

In previous lessons, students learned to solve problems in which "time = time," where time is expressed as distance/rate. Such problems involved round trips, as well as the speed of objects with and against river currents or wind speed. These problems involved one variable. Now, we come to two-variable problems in which both wind/current, and the speed of objects are to be found.

A typical problem is: "A motorboat went 18 miles downstream in 1 hour. The return trip took 3 hours. Find the rate of the boat in still water and the speed of the current."

Unlike the one-variable problems which are solved as "time = time," two-variable problems are solved as "distance = distance". The two-variable problems can also be solved as "time = time" which connect to the "distance = distance" form. I believe the connection may be a distraction for students so I choose not to present it. I did have a student who had made such connection herself. For those teachers who may have such students and/or want to present the connection as an introduction, I include the discussion here.

For the above problem, let r = speed in still water and c = speed of the current.

We write two equations: time downstream = 1 hr, and time upstream = 3 hr, where time is expressed as distance/rate.

$$\frac{18}{r+c} = 1 \textit{ and } \frac{18}{r-c} = 3$$

We multiply both sides by the respective denominators to obtain:

$$(r + c) = 18 \textit{ and } 3(r - c) = 18$$

This is now in the "distance = distance" form.

WARM-UPS

1. Factor. $(324y^2 - 289x^2)$. *Answer* $(18y - 17x)(18y + 17x)$.
2. A pilot flew a plane 480 miles with the wind. For the same length of time he then flew against the wind, but only went 420 miles. The speed of the plane in still air was 300 mph in both cases. What was the speed of the wind? (*Hint*: This is a "with/against the wind" problem). *Answer*: *Let* w = *speed of the wind*; $\frac{480}{300+w} = \frac{420}{300-w} \rightarrow 480(300 - w) = 420(300 + w)$;

   Divide both sides by 60: $8(300 - w) = 7(300 + w) \rightarrow 2400 - 8w = 2100 + 7w$;

   $15w = 300$; $w = 20$ *mph.*
3. Simplify and write with positive exponents. $\frac{(-2x^{-2})^3}{(2x^7)^2}$. *Answer*: $\frac{-8x^{-6}}{2x^{14}} = \frac{2}{x^{20}}$.
4. Solve the inequality. $|2x + 7| < 5$. *Answer*: $2x + 7 < 5$; $2x < -2$; $x < -1$

   $2x + 7 > -5$; $2x > -12$; $x > -6$; $-6 < x < -1$.
5. Which one of these points is NOT a solution of the inequality

   $2x + 5y < 5$? $(-3,-4)$ or $(-4,6)$. *Answer*: $(-4,6)$; $2(-4) + 5(6) = 22$

   $22 > 5$ *not* < 5.

Problem 1 requires students to know the squares from 11 through 20. Problem 2 is a "time = time" problem which students have solved before and segues to the lesson that follows. Problem 3 requires students to remember how to work with negative exponents as well as the rules for the power of a product. Problem 4 is an absolute value inequality with which students should be familiar. Problem 5 requires plugging in the values to see which one does not make the inequality true.

TWO VARIABLE WIND/CURRENT PROBLEMS

Problem 2 of the Warm-Ups gave us the speed of the airplane, and asked to find the speed of the wind.

"Today, we'll look at similar problems, where we do not know the speed of the wind or current, or the speed of the plane, boat, or whatever. For such problems, we use 'distance = distance.'"

> *An airplane takes 3 hours flying 1,200 miles into the wind. With the same wind, the return trip takes 2 hours. What is the speed of the wind and the speed of the plane in still air?*

"It may be helpful to use the rate × time = distance table we've used before. Help me fill it in. We don't know the speed of the plane in still air. Let's call that r. And we don't know the speed of the wind, so we'll call that w. How would we represent the downwind speed of the plane? How about upwind?"

Students may be hesitant, in which case a prompt may help: "If the speed of the plane were 300 mph, and we didn't know the wind speed, how would I represent the speed of the plane?"

Response: $300 + w$.

"And if I knew neither?" I hear "$r + w$" and they know by now that upwind speed is $r - w$. We fill in the table:

	rate	**time**	**distance**
Downwind	$r + w$	2	1200
Upwind	$r - w$	3	1200

"Since distance equals rate × time, I can write two equations."

$$2(r + w) = 1200; \textit{ and } 3(r - w) = 1200$$

"The time is multiplied by the rate on the left-hand side, and the distance of 1,200 is on the right-hand side. For some problems, like this one, we can simplify so we're not dealing with such big numbers. Both the 2 and the 3 go into 1,200, so what can I do to simplify?"

Students see they can divide both sides in the two equations by 2 and 3, respectively, obtaining: $r + w = 600$; $r - w = 400$.

They solve it. $r = 500$ mph, and $w = 100$ mph.

Examples. I work through the next one with them.

1. A motorboat can go 12 miles downstream in 2 hours. The return trip takes 3 hours. Find the speed of the boat in still water. (It's only asking for the speed of the boat, but students will still need to solve for both.)

   Answer: Let r *= speed of the boat in still water;* c *= speed of current.*

   $2(r + c)= 12$; $3(r - c) = 12$. *Divide the first equation by* 2 *on both sides and the second equation by* 3 *on both sides.* $r + c = 6$; $r - c = 4$; $2r = 10$, $r = 5$ *mph.*

   Substitute $r = 5$ *in one of the two equations;* $c = 1$ *mph.*

2. An airplane can go 1,560 km in 5 hours flying into the wind. With the same wind, the return trip takes 4 hours. What is the speed of the wind? *Answer:* $5(r - w) = 1560$; $4(r + w) = 1560$. *Divide the first equation by* 5. *Divide the second by* 4: $r - w = 312$; $r + w = 390$; $2r = 702$; $r = 351$ *mph. Substitute* $r = 351$ *in either equation to find* $w = 39$ *mph.*

I give more examples as necessary to ensure students are comfortable with solving the problems.

More Complex Problems

"Let's deal with problems where we have time expressed as a fraction of an hour."

A plane flew 180 miles in half an hour with a tail wind. With no change in the wind speed, the return trip took 40 minutes. Find the speed of the wind and plane's rate in still air.

"The first time is easy; how would you write the second time, 40 minutes as a fraction?"

Most students get these right: $\frac{1}{2}(r + w) = 180$; $\frac{2}{3}(r - w) = 180$.

Although sometimes I have to remind them, students recall that both sides are multiplied by the reciprocal of the fractions: $r + w = 360$; $r - w = 270$.

The answers are $r = 315$ mph; $w = 45$ mph.

Examples. I ask students to do some of the homework problems to serve as examples.

1. A motorboat covers 6 miles in 45 minutes. The return trip takes $1\frac{1}{2}$ hours. Find the boat's speed in still water.

Prompt: "The problem doesn't indicate which time is downstream or upstream. Would the downstream trip be shorter or longer than the upstream trip if it's the same distance?"

Answer: $\frac{3}{4}(r + c) = 6$; $\frac{3}{2}(r - c) = 6$; *multiply each side by the reciprocal.* $r + c = 8$; $r - c = 4$; $2r = 12$ *mph*; $r = 6$, $c = 2$ *mph.*

2. Larry took 36 minutes to row 3 miles. When he returned, he took 90 minutes. Find the current and speed of the boat in still water. Answer: Downstream is shorter time;

$\frac{36}{60} = \frac{3}{5}$ and $\frac{90}{60} = \frac{3}{2}$; $\frac{3}{5}(r + c) = 3$; $\frac{3}{2}(r - c) = 3$; $r + c = 5$; $r - c = 2$; $2r = 7$

$r = 3.5$ *mph*, $c = 1.5$.

3. The current in a river is 3 mph. A boat goes downstream a certain distance in 2 hours. It takes 3 hours to return the same distance against the current. Find the speed of the boat in still water and the distance the boat does downstream.

Unlike other problems so far, distance is not given. Prompt: "Let's let d = distance downstream, and r = speed of the boat in still water. Looking at the equations you've set up for the other problems, write the equations for this one." *Answer*: $2(r + 3) = d$; $3(r - 3) = d$.

Prompt: "Since the two equations both equal d, can we use substitution?"

$2r + 6 = 3r - 9$; $r = 15$ *mph*. Substitute $r = 15$ in either equation to solve for d.

$d = 2(18) = 36$ *miles*.

Homework

The problems should be a mix of two-variable and one-variable wind/current type problems. For the two-variable problems, some should ask the students to solve for distance, as in Problem 3 above.

21. QUADRATIC EQUATIONS: SQUARE ROOT METHOD

Quadratic equations for me are the pinnacle of the algebra 1 course. Typically, they are taught last, at a time when the end of school is near and students view every new topic like so much extra paraphernalia to cram into an overstuffed suitcase. That is why I make every effort to teach the unit on quadratic equations no later than April of the school year, when learning new things still matters.

Prior to teaching this unit, and after the unit on systems of equations, students learn about radicals—an essential topic for entering the world of quadratic equations. Students learn the basic properties of square roots:

Squaring and square roots of radicals: $\left(\sqrt{37}\right)^2 = 37$; $\sqrt{x^4} = x^2$

Multiplication: $\sqrt{x \cdot y} = \sqrt{x} \cdot \sqrt{y}$; $\sqrt{3} \cdot \sqrt{5} = \sqrt{15}$

Simplification: $\sqrt{128} = \sqrt{64 \cdot 2} = \sqrt{64} \cdot \sqrt{2} = 8\sqrt{2}$

Addition/subtraction: $\sqrt{128} + \sqrt{18} = 8\sqrt{2} + 3\sqrt{2} = 11\sqrt{2}$

Rationalizing denominators: $\frac{3}{\sqrt{2}} = \frac{3}{\sqrt{2}} \cdot \frac{\sqrt{2}}{\sqrt{2}} = \frac{3\sqrt{2}}{2}$

Fractional exponents: $\sqrt{x} = x^{\frac{1}{2}}$; $\sqrt[n]{x^m} = x^{\frac{m}{n}}$

Radical equations: $\sqrt{5y-1} - 8 = -1$;

$\sqrt{5y-1} = 7$; $\left(\sqrt{5y-1}\right)^2 = 7^2$; $5y - 1 = 49$; $5y = 50$; $y = 10$

In this unit, students may become confused over the various methods shown to solve quadratic equations. I like to emphasize that if an equation can be solved by factoring, then students should do so. Ultimately, they will learn that all equations can be solved by using the quadratic formula. For some equations, however, factoring is the easiest way. For example, the following equations are easier solved using factoring than using the quadratic formula: $x^2 - 5x = 0$; $x^2 - 36 = 0$.

Warm-Ups

1. $64^{\frac{2}{3}} = ?$ *Answer*: $\left(\sqrt[3]{64}\right)^2 = 4^2 = 16$.
2. Simplify. $\sqrt{243}$. *Answer*: $\sqrt{3 \cdot 81} = 9\sqrt{3}$.
3. Multiply. $x^4(-3x^{-2}y^2)^2$. *Answer*: $x^4(9x^{-4}y^4) = 9y^4$.
4. Solve. $(4x^2 = 81)$. *Answer*: $4x^2 - 81 = 0$; $(2x - 9)(2x + 9) = 0$; $x = \pm\frac{9}{2}$
5. Solve. $2x - 3 = x^2$. *Answer*: *Put in standard form*: $x^2 - 2x + 3 = 0$; $(x - 3)(x + 1) = 0$; $x = 3, -1$.

Problem 1 requires students to understand what a fractional exponent represents, and how to evaluate numbers raised to a fractional exponent. Problem 2 requires students to be able to simplify radicals by determining factors of the number under the radical that are perfect squares. In Problem 4, some students need to be reminded to put the equation in standard form (i.e., right side is zero). Problems 4 and 5 lead into this lesson; students must be familiar with factoring as a means to solve quadratic equations.

Square Root Method of Solving Quadratic Equations

After going over the Warm-Ups, I remind students that Problems 4 and 5 are quadratic equations that are solved by factoring.

"But sometimes you will have equations that cannot be factored. In this unit, you will learn how to solve any and all quadratic equations, whether factorable or not. You will be using a formula that looks like this," I say and write the quadratic formula on the board:

$$x = \frac{-b \pm \sqrt{b^2 - 4ac}}{2a}$$

This is the first time students have seen anything like this, and even though they have seen the plus or minus sign, the formula is formidable to them. I have to say that I take no small pleasure in the reactions that students evince, including "No way!" and "What are you trying to do to us?" to name a few.

"Believe it or not, when we get to this lesson, the formula will not seem as onerous as it does to you now. So, let's take this one step at a time. Just to see if you remember, who can tell me what the square root of x^2 is?"

Response: x.

Moving on, I write on the board the standard form for a quadratic equation.

$$ax^2 + bx + c = 0$$

"In Problem 4 of today's Warm-Ups, you put the equation in standard from. When you did so, you had an equation where b equals zero. You solved it by factoring the difference of two squares. But now that you've had experience with square roots, we can look at this problem another way. Rather than putting it in standard form, we can take the square root of both sides."

$$\sqrt{4x^2} = \pm\sqrt{81}$$

"We know what the square root of $4x^2$ is, I hope. Someone tell me."

Response: $2x$.

"Good. Now, what's the square root of 81? And you'll notice I put the plus or minus sign in front of the radical. This is because we want to know both roots of 81."

I want to hear 9 and –9, which I generally hear. "The solution is $\pm\frac{9}{2}$. That is, $\frac{9}{2}$ and $-\frac{9}{2}$. This is the same answer you obtained by factoring. I hear you wondering why you can't just factor it like you did for the Warm-Ups. You can. But if I asked you to solve this next equation, you'd have a harder time with factoring."

$$x^2 = 12$$

"We take the square root of each side, and we get this."

$$x = \pm\sqrt{12}$$

"Who can simplify this radical?" I wait for hands to go up; 12 is relatively easy to simplify: $x = \pm 2\sqrt{3}$.

Many students forget the plus or minus sign. One way around this is to write two answers, one with a negative sign in front and the other not. But I want them to get used to the plus or minus sign.

"In solving this problem, you used the 'property of square roots of equal numbers.'"

If r and s are real numbers and $r^2 = s^2$, *then* $r = s$ *or* $r = -s$.

I emphasize the term "real numbers" in the above. "If we have a negative number under the radical, it is *not* a real number, as we discussed previously. The square root of −12 has no real number solution." I tell them that in algebra 2 they will learn about "imaginary numbers" as a solution. "But they are not part of the set of real numbers. In this course, we will only work with real numbers."

There is a somewhat collective sigh of relief at this, though there are usually some students who are genuinely curious—and even some who know about imaginary numbers because of an older sibling.

Examples

1. $r^2 = 9$. *Answer*: $r = \pm 3$.
2. $t^2 = 49$. *Answer*: $t = \pm 7$.
3. $4m^2 = 1$; *Answer*: *first method*: $2m = \pm 1$; $m = \pm\frac{1}{2}$; *second method*: $m^2 = \frac{1}{4}$; $m = \pm\sqrt{\frac{1}{4}} = \pm\frac{1}{2}$.
4. $g^2 - 5 = 0$. *Answer*: $g^2 = 5$; $g = \pm\sqrt{5}$.
5. Solve and simplify. $2g^2 - 5 = 0$. *Hint*: Isolate the g term first.
   Answer: $g^2 = \frac{5}{2}$; $g = \pm\sqrt{\frac{5}{2}} = \pm\frac{\sqrt{5}}{\sqrt{2}} \cdot \frac{\sqrt{2}}{\sqrt{2}} = \pm\frac{\sqrt{5}}{2}$.

Equations in the Form of $(x + a)^2 = c$

"The square root method can solve more complicated equations like this one."

$$(x - 2)^2 = 4$$

"What's the square root of $(x - 2)^2$?"

If there are only a few hands (and I do admonish people not to shout out the answer), I'll ask "What did we say the square root of x^2 is?" This triggers more responses, and students are able to answer my first question: $x - 2$.

"Now, we take the square root of both sides." $x - 2 = \pm 2$.

"We want to get rid of the −2 on the left-hand side, so do it the way you always do it." I will see various forms for the answer.

$$x = \pm 2 + 2, \textit{ and } x = 2 \pm 2$$

"The second way is the best way to write it; the plus or minus comes afterward. The plus or minus symbol is just a short way of writing the following."

$$x = 2 + 2, \textit{ and } x = 2 - 2$$

"So, our two answers are 4 and 0."

Examples. These consist of some of the homework problems.

1. $9(x + 3)^2 = 16$.

Answer: $(x + 3)^2 = \frac{16}{9}$; $x + 3 = \pm\sqrt{\frac{16}{9}}$; $x = -3 \pm \frac{4}{3}$; $-\frac{9}{3} + \frac{4}{3}$ *and* $-\frac{9}{3} - \frac{4}{3}$ $x = -\frac{5}{3}, -\frac{13}{3}$.

For this problem, I remind students to first divide both sides by 9.

2. $(x + 2)^2 = 7$. *Answer*: $x = -2 \pm\sqrt{7}$.

I work through this one with them so that they get used to this form. I give one or two more like this and then one with a fraction.

3. $4(x + 3)^2 = 7$; $(x + 3)^2 = \frac{7}{4}$; $x + 3 = \pm\frac{\sqrt{7}}{2}$; $x = -3 \pm\frac{\sqrt{7}}{2}$.

This answer can be left in this form or put over the common denominator of 2: $\frac{-6\pm\sqrt{7}}{2}$.

I explain that the answer can be left this way, which is called "radical form." If I wanted a numerical answer, I would allow them to use the calculator to find the square root of 7. I want them to get used to seeing answers in this form, however, since they will be seeing it often.

4. $x^2 + 2x + 1 = 4$. I ask them whether the trinomial is a perfect trinomial square, and then have them write the equation in factored form: *Answer*: $(x + 1)^2 = 4$.

$x + 1 = \pm\sqrt{4}$; $x = -1 \pm 2$; $x = -1 -2$, *and* $x = -1 + 2$; $x = -3,1$.

5. $x^2 - 10x + 25 = 9$. *Answer*: $(x - 5)^2 = 9$; $x - 5 = \pm 3$; $x = 5 - 3, 5 + 3$, $x = 2,8$.
6. $(x - 5)^2 = 27$. *Answer*: $x - 5 = \pm\sqrt{27}$; $x - 5 = \pm 3\sqrt{3}$; $x = 5 \pm 3\sqrt{3}$.

HOMEWORK

The homework is a continuation of these types of problems. Some of them are like Problem 5 above, so students need to be able to recognize perfect trinomial squares. Answers may be left in radical form.

22. COMPLETING THE SQUARE

This lesson builds on the square root method in which equations in the form $(ax + b)^2 = c$ are solved by taking the square root of each side. Students learn to "complete the square," in which we can find a number c, which when added to an expression such as $ax^2 + bx$, creates a perfect trinomial square.

I require students to write the perfect square trinomial in factored form; that is, $x^2 - 6x + 9$ written as $(x - 3)^2$. Completing the square represents new information to be absorbed and mastered. It ultimately allows for solving unfactorable equations. I have found that application of completing the square to solving equations is usually too much information for one lesson. This lesson therefore addresses only completing the square. Solving equations by completing the square is presented in the following lesson.

Warm-Ups

1. Solve using the square root method. $(x - 3)^2 = 25$. *Answer*: $x - 3 = \pm 5$. $x = 3 + 5, x = 3 - 5; x = 8, -2$.
2. Factor. $m^2 - 4m + 4$. *Answer*: $(m - 2)^2$ or $(m - 2)(m - 2)$.
3. Solve and simplify the radical. $x^2 - 45 = 0$. *Answer*: $x = \pm\sqrt{45}; x = 3\sqrt{5}$.
4. Square the binomial. $(x + a)^2$. *Answer*: $x^2 + 2ax + a^2$.
5. Square the binomial. $\left(x - \frac{1}{x}\right)^2$. *Answer*: $x^2 - 2\left(\frac{1}{x}\right)(x) + \frac{1}{x^2} = x^2 - 2 + \frac{1}{x^2}$.

Problem 2 requires students to identify a perfect trinomial square. Also, I make a point of showing that the factored form can be a binomial squared, which is part of the day's lesson. Problem 3 requires the square root method as well as simplifying the radical. Problems 4 and 5 review the procedure for squaring a binomial, which hopefully they know how to do without resorting to actually multiplying the two binomials. The problems serve to lead into the lesson on completing the square.

UNFACTORABLE TRINOMIALS

Before my students forget the previous lesson, I refresh their memories of what we did.

"You may recall that yesterday we had some problems that had a trinomial square as one member and a nonnegative number on the other side: For example: $(x + 1)^2 = 9$.

"You then took the square root of each side. You also had some problems where the left-hand side was a perfect square trinomial."

$$y^2 + 10y + 25 = 9$$

"How did you solve it?"

Someone will tell me the process.

$$(y + 5)^2 = 9; (y + 5) = \pm 3; y = -5 \pm 3: = -2, -8$$

"Now, if you didn't happen to recognize it was a perfect square trinomial, you might have put it in standard form, by subtracting 9 from each side, like this."

$$y^2 + 10y + 16 = 0$$

"Can this be solved by factoring?" I call on someone to do the problem at the board.

$$(y + 8)(y + 2) = 0; y = -8, -2$$

"This problem can be solved either way: by the square root method or by factoring. Some of you may be wondering if you need to use the square root method for some problems, and factoring for others. No, you can use whatever method works. The square root method is a good method for problems where you can't factor, as we saw for Problem 3 of the Warm-Ups.

"But sometimes we have a problem that is not in the form where we can use the square root method. For example, this problem."

$$x^2 + 2x - 7 = 0$$

"Can this be factored?" There is general consensus that it cannot.

"Right. And it isn't a perfect square trinomial. If only it were; then, we *could* solve it using the square root method. But alas, it is not," I say with a wistful sigh.

"But wait! There is a way to make this into a perfect square trinomial. First, let's add 7 to both sides." I ask them to write this in their notebooks.

$$x^2 + 2x \quad = 7$$

"You'll notice I left some space between the $2x$ and the equals sign. There's a reason for that. Let's add one to each side."

$$x^2 + 2x = 7 + 1$$

"Anyone recognize the left-hand side?"

I'm counting on someone to recognize that it's a perfect trinomial square. Usually one or two students do, but if not, I'm willing to draw the next card as necessary.

"So, we can write the equation like this." I write the equation on the board: $(x + 1)^2 = 8$.

"Now, we *can* use the square root method, which I'd like you to do." I walk around checking notebooks, and I pick a student to write the solution on the board.

$$x + 1 = \pm\sqrt{8};\ x = -1 \pm 2\sqrt{2}$$

Completing the Square

"You may be wondering how I knew to add 1 to each side. I did it through a procedure called 'completing the square,' which you will now learn."

"I'm going to write the squares of some binomials. The squares of binomials are perfect square binomials, and there will be a pattern. I'm going to use the rule for squaring binomials, which is the square of the first term, plus two times the product of the two terms, plus the square of the last." After the first one, I ask students to tell me the steps to follow.

$$(x + 5)^2 = x^2 + 2(5)x + (5)^2 = x^2 + 10x + 25$$

$$(x - 3)^2 = x^2 + 2(-3)x + (-3)^2 = x^2 - 6x + 9$$

$$(x - n)^2 = x^2 + 2(-n)x + (-n)^2 = x^2 - 2nx + n^2$$

"You may have noticed the pattern here. In each case, the third term is the square of half the coefficient of the middle term—which is called the 'linear term.' For example, in the first one, the middle term is 2 × 5, commonly known as 10. Half of 10 is…?"

Response: 5.

"And five squared is…"

Response: 25.

"Let's write out the steps in your notebook. We complete the square for $x^2 + 6$"

$$x^2 + 6x$$

Step 1: Find half of the linear coefficient. The linear coefficient is 6: $\frac{6}{2} = 3$.

Step 2: Square the result: $3^2 = 9$.

Step 3: Add the result to $x^2 + 6x$ to get $x^2 + 6x + 9$.

"If we factor the resulting expression, we get $(x + 3)^2$. Notice that the second term is half of the linear coefficient."

Examples: I want students to complete the square and then write the result in factored form. I walk through the first one in detail.

1. $x^2 + 10x + ?$ *Answer*: $\left(\frac{10}{2}\right)^2 = 25$: $x^2 + 10x + 25 = (x + 5)^2$.
2. $x^2 - 4x + ?$ *Answer*: $\left(-\frac{4}{2}\right)^2 = 4$; $x^2 - 4x + 4 = (x - 2)^2$.

I point out that the middle term of the trinomial is negative, so the second term in the factored binomial will be negative.

3. $x^2 + 8x + ?$ *Answer*: (16); $x^2 + 8x + 16 = (x + 4)^2$.
4. $x^2 + 7x$. *Answer*: $\left(\frac{7}{2}\right)^2 = \frac{49}{4}$; $x^2 + 7x + \frac{49}{4} = \left(x + \frac{7}{2}\right)^2$.

More Complex Examples. In these, the linear term is a fraction. I remind students that to find $\frac{1}{2}$ of a fraction multiply the fraction by $\frac{1}{2}$.

1. $x^2 - \frac{1}{2}x + ?$ *Answer*: $\left(-\frac{1}{2} \cdot \frac{1}{2}\right)^2 = \left(-\frac{1}{4}\right)^2 = \frac{1}{16}$; $x^2 - \frac{1}{2}x + \frac{1}{16} = \left(x - \frac{1}{2}\right)^2$.
2. $t^2 + \frac{1}{6}t$ *Answer*: $\left(\frac{1}{6} \cdot \frac{1}{2}\right)^2 = \frac{1}{144}$; $t^2 + \frac{1}{6}t + \frac{1}{144} = \left(t + \frac{1}{12}\right)^2$.
3. $x^2 + bx + ?$ *Answer*: $\left(b \cdot \frac{1}{2}\right)^2 = \frac{b^2}{4}$; $x^2 + bx + \frac{b^2}{4} = \left(x + \frac{b}{2}\right)^2$.

This problem is challenging for most students, so I walk them through it.

Homework

Problems require completing the square, as shown in the examples. Problems include linear terms for which the coefficients of some are fractions and others are letters like the last problem above.

23. SOLVING QUADRATIC EQUATIONS BY COMPLETING THE SQUARE

Having had practice with completing the square, including those expressions where the linear term is a fraction, students are now ready to put it to use. In this lesson, students first learn how to solve such equations when the coefficient of the quadratic term equals 1. This is followed by equations where the quadratic term is other than 1. For both types of equations, solutions may involve fractional forms such as the following:

$$x = -2 \pm \frac{\sqrt{27}}{2} = \frac{-4 \pm \sqrt{27}}{2} = \frac{-4 \pm 3\sqrt{3}}{2}$$

Some solutions may also require factoring and cancelling, for example:

$$x = \frac{-2 \pm \sqrt{24}}{2} = \frac{-2 \pm 2\sqrt{6}}{2} = \frac{2(-1 \pm \sqrt{6})}{2} = -1 \pm \sqrt{6}$$

Getting used to these forms makes working with the quadratic formula easier. The quadratic formula is what this is all leading up to.

Although quadratic equations differ from linear equations, the foundational concepts and procedures in the earlier part of the course still play a role. Students must know the procedures for working with algebraic fractions, working with radicals, and factoring.

Some students find that the new information on solving quadratic equations eclipses the operations of simplification. This leaves me with some options for tests. I can penalize students by taking off points for not simplifying the final answer, even if it is technically correct. I have chosen not to do this. Instead, I indicate on the test that students can get an extra credit point for simplifying the answer. This has had the effect of motivating some (but not all) students to get into the habit.

WARM-UPS

1. Solve (all terms over one denominator and radicals simplified).

$2(x-1)^2 = 7$.

Answer: $(x-1)^2 = \frac{7}{2}$; $x - 1 = \pm\sqrt{\frac{7}{2}}$; $x = 1 \pm\sqrt{\frac{7}{2}}$; $x = 1 \pm\frac{\sqrt{14}}{2} = \frac{2\pm\sqrt{14}}{2}$.

2. How many quarts of distilled water must be added to 8 quarts of a 40% solution of nitric acid to get an 8% solution of acid?

Answer: *Let x=qts of water*; $\frac{3.2}{8+x} = 0.08$; $3.2 = .64 + 0.08x$;

$2.56 = 0.08x$; $x = 32$ *qts*. Or the equation can be multiplied by 100 to eliminate decimals:

$320 = 64 + 8x$; $8x = 256$; $x = 32$ *qt*.

3. Complete the square and factor. $x^2 - 7x + ?\ (?)^2$. *Answer*: $\left(-\frac{7}{2}\right)^2 = \frac{49}{4}$

Factored form: $\left(x - \frac{7}{2}\right)^2$.

4. Simplify. $\left(\frac{b}{a} \cdot \frac{1}{2}\right)^2$. *Answer*: $\left(\frac{b}{2a}\right)^2 = \frac{b^2}{4a^2}$.

5. Solve. $x^2 = -23 + 24x$. *Answer*: $x^2 - 24x + 23 = 0$; $(x - 23)(x - 1) = 0$

$x = 23,1$.

For Problem 1, I pick a student who has done it correctly to show the work on the board. It may help to write one as $\frac{1}{1}$ to show students see that with LCD = 2, $1 = \frac{2}{2}$, which is then combined with $\frac{\sqrt{14}}{2}$. Problem 2 is a mixture problem. Students may be confused as to whether this is a one- or two-variable problem. Prompt: "What are we solving for? One item or two?" Problem 3 requires dividing –7 by 2 to obtain $-\frac{7}{2}$. Problem 4 is straightforward multiplication and squaring. Problem 5 requires the equation to be put in standard form, which entails reordering the terms. The problem can be solved by factoring, but it will be used as an example of solving by completing the square in the lesson.

Completing the Square to Solve Equations, when *a* = 1

"What is the standard form for a quadratic equation," I ask and call on any student, whether I believe that person can answer or not. I expect all students to answer correctly at this point given how many times I've asked the question. I usually get the correct answer. If not, I'll ask for a student to provide the correction.

Standard form of quadratic equation: $ax^2 + bx + c = 0$.

"As you know, the first term is called the 'quadratic term,' the second term is the 'linear term,' and the third term, *c*, is the 'constant.' Yesterday, we learned how to complete the square. We can now use this to solve quadratic equations. We'll start out with equations where the coefficient of the quadratic term is 1. Let's start."

On the board: $x^2 + 6x + 7 = 0$.

Step 1: Add the opposite of c to both sides: $x^2 + 6x = -7$.

"Notice that I left a space after 6x. That's because we're going to complete the square."

Step 2: Complete the square: Divide the coefficient of the linear term (b) by two and square it: $\left(\frac{6}{2}\right)^2 = (3)^2 = 9$.

Step 3: Add this number to both sides: $x^2 + 6x + 9 = -7 + 9 = 2$.

Step 4: Express the left-hand trinomial as the square of the binomial: $(x + 3)^2 = 2$.

Step 5: Solve by using the square root method: $x + 3 = \pm\sqrt{2}$; $x = -3 \pm \sqrt{2}$.

Examples. I start with problems that can be solved by factoring. After solving by completing the square, I ask which method students prefer. The answer is always factoring.

1. $x^2 - 2x - 80 = 0$. *Answer*: $x^2 - 2x = 80$; $x^2 - 2x + 1 = 80 + 1$ $(x - 1)^2 = 81$; $x - 1 = \pm 9$; $x = 1 + 9, 1 - 9$; $x = 10, -8$.
2. $x^2 - x = 2$. *Answer*: *Completing the square*: $\left(-\frac{1}{2}\right)^2 = \frac{1}{4}$; $x^2 - \text{x} + \frac{1}{4} = \frac{8}{4}$; $\left(x - \frac{1}{2}\right)^2 = \frac{8}{4} + \frac{1}{4} = \frac{9}{4} \rightarrow x - \frac{1}{2} = \pm\frac{3}{2}$; $x = \frac{1}{2} + \frac{3}{2}, \frac{1}{2} - \frac{3}{2}$; $x = 2, -1$.
3. $a^2 + 12 = 10a$. *Answer*: $a^2 - 10a = -12$; $a^2 - 10a + 25 = -12 + 25$; $(a - 5)^2 = 13$; $a - 5 = \pm\sqrt{13}$; $a = 5 \pm \sqrt{13}$.

4. $a^2 + 7a + 5 = 0$. *Answer*: $a^2 + 7a = -5$. *Complete the square*:
$\left(\frac{7}{2}\right)^2 = \frac{49}{4}$; $a^2 + 7a + \frac{49}{4} = -5 + \frac{49}{4} = -\frac{20}{4} + \frac{49}{4} = \frac{29}{4}$; $\left(a + \frac{7}{2}\right)^2 = \frac{29}{4}$.

In adding $-5 + \frac{49}{4}$, I remind students they can write –5 as $-\frac{5}{1}$. This makes it easier for some to see that –5 is $-\frac{20}{4}$.

$a + \frac{7}{2} = \pm\sqrt{\frac{29}{4}}$; $a + \frac{7}{2} = \pm\sqrt{\frac{29}{4}} = \sqrt{\frac{29}{2}}$; $a = \frac{-6\pm\sqrt{29}}{2}$.

Solving Equations when a = Other than 1

"Now, let's look at equations when a is something other than 1. We divide each term by the coefficient of the quadratic term." I write on the board:

$$3x^2 - 2x - 9 = 0$$

"The first thing we do is move –9 to the other side."

$$3x^2 - 2x = 9$$

"Now, we divide every term—on both sides—by the coefficient of the quadratic term, which is 3 in this case."

$$x^2 - \frac{2}{3}x = 3$$

"We have a fraction coefficient for the linear term. So, we divide by 2 and square it." I ask a student to do this at the board.

$$\left(-\frac{2}{3} \cdot \frac{1}{2}\right)^2 = \left(-\frac{1}{3}\right)^2 = \frac{1}{9}$$

"I usually do this on the side of the page—maybe put it in a box to remind you of all your steps. Now, add it to both sides—and keep things in fraction form. So, turn 3 into a fraction."

$$x^2 - \frac{2}{3}x + \frac{1}{9} = \frac{3}{1} + \frac{1}{9} = \frac{28}{9}$$

"Now, what do we do next?"

The students know how to factor the trinomial on the left hand side, and take the square root of both sides. I ask someone to do these steps and to write the result of these actions on the board.

$$x - \frac{1}{3} = \pm\sqrt{\frac{28}{9}}$$

"What should we do with that 9 in the denominator under the radical? Finish this up in your notebooks." I go around to see what the students have done.

$$x = \frac{1}{3} \pm \frac{\sqrt{28}}{3}\text{; } \textit{or simplified}\text{: } x = \frac{1 \pm 2\sqrt{7}}{3}$$

Examples. I include some examples from the homework assignment, so we are starting on the homework. I assure the students that these problems seem difficult because there are a lot of steps, but they should just keep careful track of what they are doing. Neatness helps!

I work through the first one with the class, labeling the steps as we go:

1. $4x^2 + 12x = 11$. *Answer*: *Divide terms by 4*: $x^2 + 4x = \frac{11}{4}$.

Complete the square: $x^2 + 4x + 4 = 4 + \frac{11}{4} = \frac{16}{4} + \frac{11}{4} = \frac{27}{4}$.

Factor left side: $(x + 2)^2 = \frac{27}{4}$.

Take the square root of each side: $x + 2 = \pm\sqrt{\frac{27}{4}}$.

Solve and simplify: $x = -2 \pm \sqrt{\frac{27}{2}}$; $x = -\frac{4}{2} \pm \sqrt{\frac{27}{2}}$; $x = \frac{-4 \pm 3\sqrt{3}}{2}$.

2. $2x^2 - x = 3$. *Answer*: $x^2 - \frac{x}{2} = \frac{3}{2}$; $\left(-\frac{1}{2} \cdot \frac{1}{2}\right)^2 = \frac{1}{16}$; $x^2 - \frac{x}{2} + \frac{1}{16} = \frac{3}{2} + \frac{1}{16}$; $\left(x - \frac{7}{2}\right)^2 = \frac{25}{16}$; $\left(x - \frac{1}{4}\right) = \pm\frac{5}{4}$; $x = \frac{1}{4} \pm \frac{5}{4}$; $x = \frac{6}{4}, -\frac{4}{4}$; $x = \frac{3}{2}, -1$.

"Yes," I tell my students. "You could have solved that one by factoring."

Homework

Problems are similar to these types of problems. They include problems in which $a = 1$, and other than 1. I try to allow maximum time to work on these with the students during class time. I also will go through the more difficult ones the next day.

24. THE QUADRATIC FORMULA

I like to introduce the quadratic formula before showing its derivation, which I do the following day. I give a completing the square problem in the Warm-Ups, which I then use to demonstrate that the formula gives the same answer. I then spend time showing how the quadratic formula is used.

I also go over homework problems that involve completing the square. I make a point of showing how the radical on the right-hand side often has a denominator that is a perfect square, which serves as a shortcut. Specifically, in the day's Warm-Ups, there is a completing the square problem. At the point when they must simplify the right-hand side, I point out the shortcut:

$$x + \frac{3}{4} = \pm\sqrt{\frac{41}{16}} = \sqrt{\frac{41}{4}}$$

One of the reasons I do this is to plant the seed for the next day's derivation of the quadratic formula.

Warm-Ups

1. Multiply. $a(a^n)$. *Answer*: a^{n+1}.
2. Solve. $|x + 3| = 5$. Answer: $x + 3 = -5$; $x = -8$; $x + 3 = 5$, $x = 2$.
3. Solve by any method. $x^2 + 6x + 9 = 25$.

Answer: $x^2 + 6x - 16 = 0$; $(x - 2)(x + 8)$; $x = 2,-8$.

Or they can use the square root method. $(x + 3)^2 = 25$; $x + 3 = \pm 5$; $x = 2 - 8$.

4. A chemist mixes a 50% acid solution with an 80% solution to produce 300 mL of a 70% solution. How much of each solution do they mix?

Answer: Let x *= mL of 50% solution and* y*= mL of 80% solution.*

$x + y = 300$; $0.5x + 0.8y = 0.7(300) = 210$; $5x + 8y = 2100$;

$x = 300 - y$ *from first equation; Substituting in second equation:* $5(300 - y) + 8 = 2100$; $3y = 600$; $y = 200$, $x = 100$ *mL.*

5. Solve by completing the square. $2x^2 + 3x - 4 = 0$. *Answer: Divide terms by* 2:

$x^2 + \frac{3}{2}x - 2 = 0$; $x^2 - \frac{3}{2}x = 2$; *multiply* $\frac{3}{2}$ *by* $\frac{1}{2}$ *and square:* $\left(\frac{3}{2} \cdot \frac{1}{2}\right)^2 = \left(\frac{3}{4}\right)^2 = \frac{9}{12}$;

$x^2 + \frac{3}{2}x + \frac{9}{16} = 2 + \frac{9}{16} = \frac{32+9}{16}$; $\left(x + \frac{3}{4}\right)^2 = \frac{41}{16}$; $x + \frac{3}{4} = \sqrt{\frac{41}{16}} = \frac{\sqrt{41}}{4}$; $x = \frac{-3\pm\sqrt{41}}{4}$.

Problem 1 will cause students some difficulty. Prompts: "Can we write a with an exponent? What is the exponent?" Problem 2 is an equation not an inequality, so students can solve as shown. Students may notice that Problem 3 has the same answer. I point out that if we square both sides of the equation in Problem 2, we get Problem 3. Problem 4 is a mixture problem that they have had before. Students may need to be reminded to use substitution for solving the two equations. Problem 5 is similar to their homework problems.

THE QUADRATIC FORMULA

I go through the answer to Problem 5 of the Warm-Ups step by step. We then go through any homework problems that cause difficulty. Generally, students have trouble working with fractions, particularly combining a whole number on the right-hand side with the "c" term that is added to both sides. I advise them to write the whole number with a denominator of one to help them see that $2 = \frac{32}{16}$.

"Let's look at Problem 5 of the Warm-Ups again."

$$2x^2 + 3x - 4 = 0$$

"What is the standard form of the quadratic equation?" I ask this question often, so most students will know the answer, and I write it under Problem 5:

$$ax^2 + bx + c = 0$$

"Completing the square is used for a variety of things. One of them is solving the standard form of the quadratic equation using it. When we do that, we obtain a formula that can be used to solve any quadratic equation. It's called the quadratic formula. I showed it to you at the beginning of this unit. We will see tomorrow how it's derived."

I write the formula on the board:

$$x = \frac{-b \pm \sqrt{b^2 - 4ac}}{2a}$$

Using the Quadratic Formula

"Let's see how it works using Problem 5."

$$2x^2 + 3x - 4 = 0$$

$$ax^2 + bx + c = 0$$

Step 1: Put the equation in standard form. "It's already in standard form," I say.

Step 2: Identify a, b, and c.

We want to match up the a, b, and c values of the standard from with this equation. "What is a in this equation?" Students see it is 2; I write $a = 2$ on the board.

"How about b and c?" Students identify these readily, although some will say c is four. "The sign stays with the variable and the number," I say, and I write the values on the board as well: $b = 3$, $c = -4$.

Step 3: Substitute the values of a,b, and c into the quadratic formula and evaluate.

"Now, we plug them into the formula. What is $-b$?" Students tell me it is -3, and I begin substituting the values.

$$x = \frac{-3 \pm \sqrt{3^2 - (4)(-4)(2)}}{4}$$

$$x = \frac{-3 \pm \sqrt{9 - (4)(-8)}}{4}$$

$$x = \frac{-3 \pm \sqrt{9 - (-32)}}{4}$$

I point out that we are subtracting a negative. "What happens when we subtract a negative number?"

Response: You add.

$$x = \frac{-3 \pm \sqrt{41}}{4}$$

"This is the same answer we obtained by completing the square. Which do you think is easier?"

Most will say the formula. Some will ask why they weren't taught this strategy in the beginning.

"Good question," I say. "I want you to see how completing the square leads to this formula, which we will do tomorrow. But another reason is that you will use completing the square for other problems in algebra 2 and in pre-calculus."

I also emphasize that when using the formula, they need to be careful when subtracting the 4ac term. "If c is negative, you will end up adding. If c is positive, you will be subtracting. That's why I wrote it as I did with the parentheses."

Examples. The examples are taken from the homework assignment. Some of the problems can be solved by factoring, and I tell the students this. "For today's homework, you will use the quadratic formula for all problems. In the future, though, you may use whatever method you find easiest. But the formula can be used for any quadratic equation."

1. $2x^2 + 4x + 1 = 0$. *Answer*: $a = 2$, $b = 4$, $c = 1$; $x = \frac{-4\pm\sqrt{16-(4)(2)(1)}}{4} = \frac{-4\pm\sqrt{16-8}}{4} = \frac{-4\pm\sqrt{8}}{4}$.

Simplified: $\frac{-4\pm2\sqrt{2}}{4} = \frac{2(-2\pm\sqrt{2})}{4} = \frac{-2\pm\sqrt{2}}{4}$.

2. $3x^2 + 5x + 1 = 0$. *Answer*: $a = 3$, $b = 5$, $c = 1$; $x = \frac{-5\pm\sqrt{25-12}}{6} = \frac{-5\pm\sqrt{13}}{6}$.

3. $x^2 = 4x - 3$. *Answer*: $x^2 - 4x + 3 = 0$; $a = 1$, $b = -4$, $c = 3$;
   $x = \frac{-4\pm\sqrt{16-(4)(2)(1)}}{4} = \frac{-4\pm\sqrt{16-12}}{4} = \frac{-4\pm\sqrt{4}}{2} = \frac{4+2}{2}, \frac{4-2}{2} = 3,1$.

4. $u^2 = 14$.

For this answer, I ask for the equation to be put in standard form: $u^2 - 14 = 0$. "Is there a linear term?" No. "So, there's no *b*. So, we write $b = 0$. What are *a* and *c*?" 1 and −14.

$x = \frac{-0\pm\sqrt{0 - 4(1)(-14)}}{2} = \frac{\pm\sqrt{96}}{2} = \frac{\pm2\sqrt{14}}{2} = \pm\sqrt{14}$.

Students will say it was harder using the formula than using the square root method, and I agree. I say, "I want you to get used to using the formula."

5. $x^2 + 9x = 0$. *Answer*: $a = 1$, $b = 9$, $c = 0$; $x = \frac{-9\pm\sqrt{81-(4)(1)(0)}}{2} = \frac{-9\pm\sqrt{81}}{2} = \frac{-9\pm9}{2} = \frac{0}{2}, -\frac{18}{2} = 0,-9$.

Homework

The homework continues with problems of this type. In some cases, the equation will not be stated in standard form. Students must remember to put all equations in standard form in order to identify the *a*, *b*, and *c* values correctly. I also assign one or two word problems, one of which is below:

Two tin squares together have an area of 325 square inches. One square is 5 inches longer than the other. Find the length of the sides of each square.

I advise students that they will get two answers, one of which will be negative. Since a length cannot be negative, they must reject that answer.

Answer: Let x = length of side of smaller square. Then, $x + 5$ is the length of the side of the larger square. The area of a square is the length of its side squared.

$$x^2 + (x + 5)^2 = 325; x^2 + x^2 + 10x + 25 = 325; 2x^2 + 10x - 300 = 0$$

Equation can be simplified: $x^2 + 5x - 150 = 0$; $(x - 5)(x + 10) = 0$;

$x = 5, x + 5 = 10$.

Students will find it easier to factor than to use the quadratic formula. If they use the formula, they will need to know that the square root of 625 is 25.

25. DERIVATION OF THE QUADRATIC FORMULA

The derivation of the quadratic formula is accomplished by solving the standard form of a quadratic equation by completing the square. I tell students that on the next test there is an extra credit question worth 10 points, asking for the derivation of the formula.

When I was in education school to get my teaching credential I took a math teaching methods course. In that class we had to prepare a set of lesson plans for a unit. Mine was on quadratic equations, and I mentioned my idea of allowing extra credit on a test for those who could derive the formula. The teacher felt that students would merely memorize the steps and would not understand what was really going on.

Having done this for several classes, my experience is that students who try to memorize the steps may not totally succeed, but the process of memorization will bring into play what is involved with completing the square, so there is some understanding. Those who fully understand the process of completing the square and how it is used to derive the quadratic formula, usually fare better.

Today's Warm-Ups contain some problems that front-end load some of the steps that are used in the derivation.

WARM-UPS

1. Solve using the quadratic formula. $y^2 + 8y + 13 = 0$.

Answer: $x = \frac{-8 \pm \sqrt{64-52}}{2} = \frac{-8 \pm \sqrt{12}}{2}$.

Simplified: $\frac{-8 \pm 2\sqrt{3}}{2} = \frac{2(-4 \pm \sqrt{3})}{2} = -4 \pm \sqrt{3}$.

2. Complete the square. $x^2 + \frac{b}{a}x + ?$ *Answer*: $\left(\frac{b}{2a} \cdot \frac{1}{2}\right)^2 = \frac{b^2}{4a^2}$.
3. Combine. $\frac{b^2}{4a^2} - \frac{c}{a}$. *Answer*: $LCD = 4a^2$; $\frac{b^2}{4a^2} - \frac{c}{a} \cdot \frac{4a}{4a} = \frac{b^2-4ac}{4a^2}$.
4. Simplify. $\sqrt{4a^2}$. *Answer*: $\sqrt{4a^2} = \sqrt{3} \cdot \sqrt{a^2} = 2a$.
5. Solve using any method. $x^2 + 15x = 0$. *Answer*: $x(x + 15) = 0, x = 0,-15$.

Problem 1 is straightforward, though some students may have difficulty simplifying it. Problems 2–4 are front-loading procedures students will use in the day's lesson on deriving the quadratic formula. Prompt for Problem 2: "What is the coefficient of x?" "How do we find half of that?" Problem 3 is addition of fractions, since the LCD is $4a^2$. Prompt: "What do you multiply $\frac{c}{a}$ by to get $4a^2$ in the denominator. For Problem 4, students may need to be reminded that the square root of a variable to a power is the exponent of the variable divided by two. For Problem 5, most students will factor, but some will still miss that one of the solutions is $x = 0$.

By Way of Introduction

I go through the Warm-Ups carefully since three of the problems are procedures they will use in the derivation of the formula. I start off the lesson by having a student read aloud one of the posters I have on my wall—a quote from Rene Descarte:

"Each problem that I solved became a rule, which served afterwards to solve other problems."

"We've seen this all year long. You learn a rule for solving problems and then use it in more complex problems. This quote is very apt for deriving the quadratic formula, which is what we will do today. The quadratic formula is the end result of solving the standard form of the quadratic equation by completing the square. So, with that, let's get started."

Derivation of the Quadratic Formula

I write the standard form for a quadratic equation on the board: $ax^2 + bx + c = 0$.

I ask students to write this and all the steps that follow in their notebooks.

"We want to complete the square. What is usually the first thing we do?"

There are two possible answers that students give, and I'll go with either one. One answer is to move c to the other side. The other is to divide all terms by a. Assuming the first possibility is given, that's what I do; otherwise, I divide all the terms by a and then move c to the other side.

Moving c to the other side: $ax^2 + bx = -c$.

Dividing all terms by a: $x^2 + \frac{b}{a}x = -\frac{c}{a}$.

"What's next?" There's usually silence. "Well how have you done it in the past? What is this procedure called?" This usually jars the answer loose: Complete the square.

"What is the coefficient of the linear term?"

Response (with some hesitation): $\frac{b}{a}$.

"Now, actually you've already done this in your Warm-Ups, Problem 2. So, tell me what it is I need to do."

Someone will tell me to multiply $\frac{b}{a}$ by $\frac{1}{2}$ and square it.

$$\left(\frac{b}{a} \cdot \frac{1}{2}\right)^2 = \frac{b^2}{4a^2}$$

"Now, what do I do with this?"

Response: Add it to both sides.

$$\textit{Completing the square: } x^2 + \frac{b}{a}x = \frac{b^2}{4a^2} = \frac{b^2}{4a^2} - \frac{c}{a}.$$

"I wrote the right-hand side in the order I did because it will make things a bit easier in the end. So, what do we do now that we have a perfect square trinomial on the left-hand side?"

Someone will know that we factor it. I may need to remind them that the second term of the factored form is the square root of the third term of the perfect square trinomial.

$$\textit{Factoring the left hand side: } \left(x + \frac{b}{2a}\right)^2 = \frac{b^2}{4a^2} - \frac{c}{a}.$$

"Things are starting to look a little familiar." The students agree. "Now, let's make the right-hand side a bit more presentable. We have the sum of two fractions. We need to add them, and guess what? You already did that in Problem 3 of your Warm-Ups. Tell me what to do. In fact, come on up here and do it." I pick a student who I know won't mess it up—it's important not to break the flow here.

Making a single fraction on right hand side: $\frac{b^2}{4a^2} - \frac{c}{a} \cdot \frac{4a}{4a} = \frac{b^2 - 4ac}{4a^2}$

"Now, our equation looks even more familiar."

$$\left(x + \frac{b}{2a}\right)^2 = \frac{b^2 - 4ac}{4a^2}$$

"What is the next step?" I do this one because I don't want them to do two steps in one.

Take the square root of each side: $\left(x + \frac{b}{2a}\right) = \pm\sqrt{\frac{b^2-4ac}{4a^2}}$.

"This is getting exciting now. Can anyone simplify the right-hand side, and yes, you did that in your Warm-Ups. I think I made this too easy for you."

Someone will tell me to make the denominator $2a$ since it's the square root of $4a^2$.

Simplify the right hand side. $x + \dfrac{b}{2a} = \pm\dfrac{\sqrt{b^2 - 4ac}}{2a}$.

"Now, we solve for *x*."

Solving for x: $x = \dfrac{-b}{2a} \pm \dfrac{\sqrt{b^2 - 4ac}}{2a}$.

"Last step: make it one fraction."

Simplifying: $x = \dfrac{-b \pm \sqrt{b^2 - 4ac}}{2a}$.

"And there you have it. The quadratic formula."

Homework

I assign more quadratic equations, but this time allow them to pick the method of their choice. I also want them to work on simplifying the final answer. For example:

1. $x^2 - 2x - 8 = 0$. *Answer: Factoring is the easiest method*:

$(x - 4)(x + 2) = 0$; $x = 4, -2$.

2. $x^2 + 2x = 7$. *Answer: Quadratic formula*; $x^2 + 2x - 7 = 0$

$x = \frac{-2 \pm \sqrt{4+28}}{2} = \frac{-2 \pm \sqrt{32}}{2} = \frac{-2 \pm 4\sqrt{2}}{2} = \frac{2(-1 \pm 2\sqrt{2})}{2} = -1 \pm 2\sqrt{2}$.

Some word problems are also included, for which I provide guidance.

3. A rectangle is 5 feet longer than it is wide. The area of the rectangle is 66 square feet. Find its dimensions.

Answer: *Let* x = *width*; *length* = $x + 5$; $x(x + 5) = 66$;

$x^2 + 5x - 66 = 0$; $(x + 11)(x - 6) = 0$; $x = 6$ *ft*, $x + 5 = 11$ *ft*. (*The answer of* $x = -11$ *is rejected because all lengths must be positive.*)

4. A rectangular floor of 147 square feet is 3 times as long as it is wide. It is divided into a rectangle twice as long as it is wide and a square. Find the dimensions of the rectangle and the square.

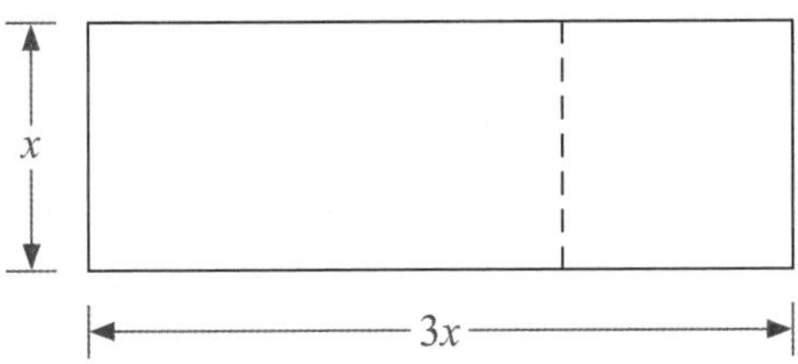

Answer: *Let* x = *width of rectangle*; *side of square* = x;

area of square + *area of smaller rectangle* = 147.

x = *length of side of square, length of smaller rectangle is* $3x - x = 2x$.

$x^2 + 2x^2 = 147$; $3x^2 = 147$; $x^2 = 49$; $x = 7$ *ft*. *Rectangle width* = $x = 7$ *ft*.

Rectangle length = $2x = 14$ *ft*.

References for Parts II and III

Brown, R. G., Smith, G., & Dolciani, M. (1988). *Basic algebra*. Houghton Mifflin.

Foerster, P. A. (1990). *Algebra 1: Expressions, equations and applications* (2nd ed.). Addison-Wesley.

Dolciani, M. P., Berman, S. L., & Freilich, J. (1962). *Modern Algebra : Structure and method*. Houghton Mifflin.

JUMP Math. (2015). *JUMP Math Assessment and Practice 7.1 and 7.2: Common Core edition*. Toronto

Singapore Math. (2003). *Primary mathematics* (U.S. ed.; textbooks 5A, 5B, 6A, 6B). Marshall Cavendish Education.

AFTERWORD

Thank you for reading this glimpse of how to teach math in a traditional manner. It is a glimpse based on our two teaching styles as well as our observations of other teachers. As I was writing the lesson examples for seventh and eighth grade, I would show it to teachers I knew to get their opinions on various matters. I often heard, "That's exactly how I do it." This is not to say that what we have written embodies the manner and technique of all traditional-minded math teachers. Everyone has their own particular style and approach. Nevertheless, we feel that we have recreated the look and feel of explicit teaching in the classroom, as well as the typical reactions of students.

One event comes to mind as I think about traditional math. It is a memory of my eighth-grade algebra class at the first school where I taught. During Indian Summer (September and October) my classroom became a hothouse in the afternoon when I taught that class. In addition to a fan at the front of the room, I thought to bring in several cartons of bottled water.

Passing the waters out to the class became a daily routine. As I distributed the waters, I would say, "I am haunted by waters," after which I would inform the class that this utterance was the last line of *A River Runs Through It*, a book by Norman Maclean, which was made into a movie. Only one student had heard of the movie.

I continued distributing the waters even as temperatures cooled, as well as saying the quote, to which the class would always respond with "River Runs Through It!" as if we were practicing a mathematical procedure. I hoped that this ritual would lead to someone interested enough to read the book, of which I kept a copy in my classroom. No one ever did, but

I maintained my faith that some of my students would read it some day, just as I believe that mathematical procedures often lead to a deeper understanding.

In an interview with Normal Maclean about his book, he talks about the art of fly-fishing and how he presents it in his book. He makes a point, which I believe is at the heart of how J.R. and I structure instruction:

> *One principle of progression in the story is the order in which you learn about the art of fly-fishing.... Only one aspect of the art of fly-fishing is presented each time you see [my brother] fish—the human mind does well to absorb one technical aspect of art at a time.*

J.R. and I agree this rings true for learning the art of mathematics. We know what it is like to be continually confronted with new information, and the effort it takes to keep things straight. Our hats are off not only to our students who persevered, but also to the teachers who maintain the structures of a traditional classroom in the face of forces advising otherwise .